OWN

THE

ARENA

Getting Ahead, Making a Difference,
and Succeeding as the Only One

OWN

THE

ARENA

Katrina M. Adams

AMISTAD

An Imprint of HarperCollins*Publishers*

HarperCollins books may be purchased for educational, business, or sales promotional use. For information, please email the Special Markets Department at SPsales@harpercollins.com.

FIRST EDITION

Designed by SBI Book Arts, LLC

Library of Congress Cataloging-in-Publication Data is available upon request.

ISBN 978-0-06-293682-0

21 22 23 24 25   LSC   10 9 8 7 6 5 4 3 2 1

In loving memory of my mother and father,

Yvonne and James Adams.

CONTENTS

12 MATCH POINTS

FOR WAYS TO THRIVE WHEN YOU'RE THE ONLY ONE

1. *Own the table:* Recognize that you are at the table for a reason. It is no accident.

2. *Own your legacy:* Set an example from beginning to end because your accomplishments will impact those who follow.

3. *Own your courage:* Be bold in all things. You stand on the fearless shoulders of those who came before you. Allow their example to lift you higher than most would dare to go.

4. *Own your identity:* Celebrate your own background and embrace the backgrounds of others.

5. *Own your choices:* The challenge is to let go of the personal sacrifices and trust that your decision is the right one.

6. ***Own your network:*** Ask yourself, *Who can help me advance? Who can help me be my best self?* Surround yourself with those individuals.

7. ***Own your village:*** Surround yourself with family, friends, and chosen confidants who you trust, to protect and empower you throughout your journey.

8. ***Own your voice:*** Do not fear speaking out. Bring your knowledgeable point of view to the conversation.

9. ***Own your successes:*** Acknowledge and celebrate your accomplishments. Humility is overrated.

10. ***Own your losses:*** Learn from your setbacks. Your losses are never truly failures. Own them because they add to your unique set of experiences and wisdom.

11. ***Own your obligation:*** If you are the first, you want to be the first of many. Reach back to pull forward.

12. ***Own the possibilities:*** Be open to a new path and don't be afraid of the curve. Because when you explore the possibilities for success, your options are limitless.

OWN

THE

ARENA

1

CENTER COURT

You either walk inside your story and own it or you stand outside your story and hustle for your worthiness.

—BRENÉ BROWN

It was one of those steamy summer evenings in New York City, where the jagged Manhattan skyline viewed from Queens appeared to soften in the hazy distance. The air was so still that, in between the squeaking of rubber tennis shoes, the grunts of the players, and the bouncing of their tennis balls, I could hear moths buzzing kamikaze-style up against the lights of Arthur Ashe Stadium. Or was it the buzz of excitement from the people in the stands, bathed in the glow of what was, at that moment, the center of the tennis universe?

The 2018 US Open took place during a record-hot week, with temperatures nearing 100 degrees Fahrenheit. For the first time in our history, we had to implement an extreme-heat policy, allowing men to take a ten-minute break between the third and fourth sets if a player made the request, and women were entitled to take a ten-minute break after the second set. We went through thousands of towels, and players were guzzling as much as two and a half gallons of water each during matches. Just walking around the grounds made me feel like I was going to melt into the asphalt. I could just imagine the physical toll felt by our players. Having played in Melbourne, Australia, and "Hotlanta" myself, I knew what they were going through.

On the night of the women's finals, the President's Suite was packed to capacity with VIPs, from well-heeled corporate titans to celebrities, including Robin Roberts, Taraji P. Henson, Shonda Rhimes, Judy Smith, Mayor David Dinkins, and Dr. Mehmet Oz. Getting this coveted invitation takes months of delicate negotiations that can become politically fraught, because everyone wants a spot, especially during the finals. But our guests must earn the privilege of being there. Each year the US Tennis Association (USTA) uses the opportunity to recognize our biggest supporters, individuals who have promoted the sport in various ways as donors, influencers, and fans.

I also exercised my prerogative as the first African American chairman and president, to invite family, friends, and business partners of color, who brought more flavor into a historically White suite. In America, we still primarily live in segregated social circles and people naturally gravitate toward others who look like them. The USTA has had only nine black directors on its board, including me, in its 138 years of existence. Without hesitation, under my lead, the President's Suite was going to reflect my priority to elevate diversity and inclusion.

Sitting with me in the front row was tennis great Virginia Wade and executive advisor Roberta Graves, and my girls from Northwestern were in the surrounding seats. I'd been having a blast watching two of my favorite players—Serena Williams and Naomi

Osaka—the GOAT (greatest of all time) and the Future—battle it out on the center court.

I'd had my eye on Naomi for a while. Her performance throughout that entire 2018 US Open had been extraordinary. Even more outstanding was her laser focus and a poise that was well beyond her years. Off court, her shy demeanor belied a fierce competitiveness. It seemed like nothing could ruffle this twenty-year-old phenomenon. Of Haitian and Japanese descent, Naomi was born in Japan and grew up mostly in the US, although it was decided early on by her parents that she would represent Japan as a player. If she won this match, it would make her the first top-ranked Asian player in the history of the sport. She was also a black woman playing against another black woman who had made history. The pressure Naomi felt must have been immense. But she has learned how to tune out the distractions and be fully in the game. This match, against an icon of the sport whom she had admired her whole life, was no different. I could see that Naomi was in a zone. With just moments to go in the match, I left my seat to prepare for the trophy presentation.

Although anything can happen in the final few games of a match, Naomi was leading and was theoretically two games away from winning, which meant it was time to make my way down to the court, to perhaps crown a new Grand Slam tournament champion or sit there for a long comeback and crown a champion of a record-tying twenty-four Grand Slam tournament titles holder. The score was 4-3 in the second set and Naomi had just broken Serena Williams's serve.

As chair and president of the US Tennis Association, one of my favorite parts of the job was to stand on the podium and present the trophies to the winner and runner-up—a ceremony that takes place almost as soon as the players have shaken hands over the net. Having been a player and dreamed of winning a major myself, this was the closest I got to ever holding the coveted champion trophy. The setting up of the stage, which for the sake of the global television broadcast gets done within ten to fifteen minutes and is timed to the second, happens right in front of the net. Most viewers don't see this well-coordinated process as a team of about ten workers swiftly assemble the pieces of the podium while the USTA's half-dozen dignitaries quietly file along the sideline. Instead, the cameras are zeroed in on the winner leaping into the stands to embrace his or her family and team members, as well as the faces of the players as they're toweling off and collecting themselves after all the drama of victory and defeat.

Knowing the finely tuned logistics of what was about to take place, I had to hustle down to center court. It would be a long walk from my seat, up the stairs, and across the carpeted suite. Still in the suite, I passed the television monitors, the empty champagne and honey deuce melon cocktail highball glasses stacked on the bar and side tables. The fifteen-hundred-square-foot space was unusually quiet. The few people who weren't outside, transfixed by the match, had their eyes riveted on the video monitors. I glanced up quickly and the score still hadn't changed.

I remembered what it was like to be in Naomi's position. Early on in my career, in 1988, I found myself in the fourth round at Wimbledon playing against Chris Evert, one of the all-time great women players. I was just nineteen, three majors into my career, and until that week I had never won a singles match at a Grand Slam tournament, losing first round at the Australian Open and French Open. We were on Court No. 2, which was the third-best court after Center and Court No. 1, although it was called the "graveyard" because it was where many of the top seeds had unexpectedly lost their matches over the years. The fact that I was half Chrissie's age was also in my favor. But she was every young female player's idol. She was not only an amazing athlete but also brought her own strong but feminine flair to the court. I'd be going out against someone I'd watched my entire career. Like so many other female players of my generation, I looked up to her. My first racket, by Wilson, even bore her autograph!

I started strong and was up 5-3, serving and volleying, chipping and charging, not giving her a chance to get a rhythm. Then I made the mistake of looking at the scoreboard. Realizing I was about to win that set, I started thinking about what it would be like to beat this champion and suddenly my head was out of the game. I managed to hold on to the first set, winning it 7-5 but then, at 3-3 in the second set, I was done. I had nothing left and Chrissie being the champion she was proceeded to win the next nine games.

Chrissie was to me what Serena now was to Naomi—an opponent to be revered and feared. She was so close she could almost taste it, but she had to put that fact out of her mind. Although she had beaten Serena, earlier in the year in Miami, this was different. It was going to take all the strength, focus, and precision Naomi could muster to beat Serena, on the grandest stage in tennis.

Before continuing on my journey, I made a quick stop at the ladies' room to check my hair and makeup. Since all the world's cameras would be on me in just a few moments, I wanted to make sure I didn't have any food particles or lipstick on my teeth. Then I grabbed my phone from the charging station at the suite's front desk. I made my way along the hallway, down two flights of stairs, then through the inner corridor that would bring me into the center of the action.

Over the last four years, I'd whittled the walk down to center court to just under five minutes, but I never knew whom I might bump into. Having my phone to my ear gave me added protection from the well-meaning well-wishers, even if I wasn't actually talking to anyone. My position as head of the USTA was not unlike being the mayor of a small town. When folks see you walking down the main street, just about everyone has something to say, from the eager ball kids to the executive sponsors whose money helped make all of this possible.

As I hurried along to center court, my train of thought jumped back and forth between past and present, caught between the myriad details to be overseen in the next days and, because this was to be my last US Open in this role, reflections on the past. It had been a long road to this moment, starting with that steamy July day in 1975 when I watched Arthur Ashe win Wimbledon on our family's grainy black-and-white television screen. I remember thinking, "You can be on TV doing this?" It was inspiring to see an African American man—the first in history to play in the Wimbledon Finals and win—take away the crown of the top-seeded player, Jimmy Connors. I told myself that could be me someday. I was six.

Now, forty-three years later, this was to be my last Women's Singles final in my last year as head of the USTA. I was getting choked up at the thought of it. Although I would always be in some way involved with the board and other national and international tennis organizations, this particular Grand Slam tournament marked the end of an incredibly rewarding period in my life and career. It was the first time we were able to showcase the completion of the $600 million, five-year transformation of the USTA Billie Jean King National Tennis Center, including the new Louis Armstrong Stadium, with a retractable roof and state-of-the-art ventilation.

I savored every moment of my last US Open, even the problems.

During a heat-mandated break, French player Alizé Cornet, returned to the court after a wardrobe change and realized her top was on backward. She retreated to the back of the court to do a quick reverse of her top, exposing her sports bra, resulting in a code violation. It was the wrong call by the chair umpire, who had not realized that was allowed from the sideline. The fact that the shirt change took place behind the baseline was what gave him pause. That said, men weren't held to the same standards and changed their shirts in full view all the time. We were swift to condemn the decision and apologized to the player, clarifying the policy going forward. Of course, it was embarrassing, but the incident also demonstrated the USTA's transparency and responsiveness to the situation—a level of communication that had been one of my priorities as chair and president.

Overall, I was beyond proud of what we'd been able to accomplish. Minority and women players had made huge strides in the sport, and although I can't take all the credit for this, I had been involved in several decisions that moved things forward.

I've always felt a responsibility to make things better for those who come after me. My mother always told me that blazing a trail isn't just a matter of getting there first; it's about being the first of many. Female players have become some of the highest paid in professional sports, and while by no means had we achieved pay parity with male players in terms of endorsement deals, American women are among the biggest tennis stars in the world. Or I should say the biggest stars. Period. Two of them—an African American and a Haitian Japanese—were on that center court where I was now headed in my three-inch heels.

I'd had my eye on Serena and her sister Venus when they first became known at the ages of nine and eleven, respectively. In 1998, during my final few years as a pro player, our paths crossed regularly. Though young, the two girls were already well into their professional careers by this time. Venus had been on tour since 1994 and Serena since 1996. They had not yet won a Grand Slam tournament singles title, but they emerged as a great doubles team, having won Mixed Doubles with their respective partners at Wimbledon. We were all playing in the European Indoors in Zurich, Switzerland, where I'd bonded with the sisters. I stopped by their hotel room one evening so that their mom, Oracene Price, could tighten a couple of my hair braid extensions. Venus and Serena sat on the bed while I sat on the floor and we started chatting about the game.

Venus and Serena's success over the subsequent years had made me proud. My friends in the President's Suite, each of whom had traveled with me occasionally on the Women's Tennis Association (WTA) Tour and gotten to know the sisters as well, felt the same. They weren't shy about whooping and cheering each time Serena scored a point against Naomi. They wanted nothing more than for me, at my last US Open Women's Final, to present the award to Serena. They especially wanted the victory for her because of the tough year Serena had and knowing the enormous pressure she had put herself under to tie Margaret Court for most Grand Slam tournament titles in the history of tennis. They also knew the kind of pressure I had endured as the only black woman in leadership of the tennis world, and they wanted to see both their friends sharing a historic moment together.

I felt Serena; I really did. But I was also a big fan of Naomi's. Over a few short years, tennis fans had seen her coming to make her mark. I saw her for the first time in Singapore a few years earlier, where she won the Rising Star event at the WTA Finals. In March 2018, Naomi beat Serena, which had made this matchup even more anticipated than usual. I'd watched Naomi rise to the occasion throughout the US Open, especially in the semifinals when she beat Madison Keys to secure a spot in her first Grand Slam tournament final. This star had now risen. Although I was disappointed for Madison, another African American woman, I was happy for Naomi, who'd really been proving herself that year. Each win made me feel like a proud mama hen. Much like her own hero, Serena, Naomi was an incredible athlete who moved with breathtaking power and speed. At just twenty years old and a little more than half her opponent's age, Naomi had a maturity that was beyond her years. Serena's success in the sport and her trailblazing as a strong female player of color had set the table for this brilliant young athlete. I couldn't wait to see what she'd do next. Whoever won that day I truly believed that the game of tennis couldn't lose. *I* couldn't lose.

Then, just as I entered the hallway that took me to the perimeter of center court, there was a deafening roar. New York crowds tend to be raucous. They're passionate about the players they support and not the least bit shy about letting their feelings be known if a point doesn't go their way. But this was different. The booing was so loud it was clear something had gone horribly wrong. There is no booing in tennis. This is a sport that has the word *love* in the score. Tennis is known as the sport of nobility, with strict rules of etiquette, or sportsmanship, that you would expect.

As rowdy as US Open fans tend to be, you only ever hear boos at a match when something egregious happens, like Ilie "Nasty" Năstase kicking over the watercooler, or when Xavier Malisse, a junior player at the time, rolled the umpire in his chair to the center of the court. There was also the time when Pablo Carreño Busta lost his cool in a spectacular fashion at the 2019 Australian Open. Up 8-5 against Japan's Kei Nishikori in the deciding tiebreaker of the fifth set, a controversial line call cost Pablo the point. In his fury he lost his focus, as well as the next three points, then the match as his opponent finished with an ace. At that point, Pablo screamed at the umpire, then hurled his racket bag before storming off the court to a chorus of boos from the crowd, although he later apologized for his unsportsmanlike exit.

I was confused as I emerged courtside. The score was 5-3 and Naomi was just a game away from taking the title. From what I had already witnessed from the suite, she had maintained her composure. Steady and calm had been her style throughout the tournament. But few people could be heard clapping for this young player, who wasn't giving the show of emotion to which fans are most likely to respond. No doubt they admired her performance, but it was clear they desperately wanted the American to win on her home court. You could feel how much the energy from the stands was with Serena. After the previous year she had endured, dealing with injuries and the fight for her life following her pregnancy, the fans desperately wanted this win for her. Their emotion was palpable. When she won a point, the place would erupt.

As I would learn later, the problems started in the second game of the second set, when chair umpire Carlos Ramos, who was known on the tour for being a no-nonsense stickler for rules, cited Serena for a hand signal (coaching) from her coach and gave her a warning under the code of conduct rules. At that point she was ahead of Naomi at 3-1 and did not receive any scoring penalties, just a warning, but the admonishment set off a cascade of contentious incidents as the match continued. She disagreed with the warning and denied that she had seen a coaching signal at all. She argued with Carlos, insisting that she had not received coaching. "I don't cheat to win; I'd rather lose," she explained. "You owe me an apology! I have never cheated in my life!"

Serena played a poor game and lost, shrinking her lead over

Naomi to 3-2, losing the break and going back on serve. After missing a shot later on in the second set, going down a service break 4-3, Serena smashed her racket onto the court, breaking it, resulting in a second code violation and the loss of a point in the next game. But Serena didn't hear that code violation over the noise from the crowd. She only discovered the ruling upon attempting to serve, when Carlos announced the score as love-15, further stoking the tension. During the next changeover, she walked to the umpire's chair and pleaded with him to tell the crowd she was not cheating.

"I didn't get coaching. You need to make an announcement that I didn't get coaching," Serena told him. "I didn't cheat. How can you say that? I have never cheated in my life. I have a daughter and I stand for what's right for her. You owe me an apology."

It was true, Serena never sought out on-court coaching on the WTA Tour and had always played with absolute integrity, even if she looked to her player's box after almost every point, which is why the umpire's call was surprising to her. As tennis fans know, all coaches offer some sort of hand signal or verbal encouragement during play, which is not allowed, but enforcing those rules, regardless of whether the player saw the coaching, seemed a little harsh. Beyond that, the interaction was a painful trigger for Serena, who had pushed past so many obstacles to be where she was now. Her passion in those final moments of the match was the culmination of all the injustices she felt subjected to in an arena that has not always been welcoming to her, from the petty slights of the French Open, where officials banned her catsuit look on the court, post-tournament, to the persistent and outrageous sexist and racist attacks in the press and social media. Plus, over the years, Serena was often on the receiving end of a bad call.

In that moment, pleading with Carlos, we could all feel her passion and frustration. If nothing else, after all that she has given to the sport, the world should know that she is the last person who would try to gain an unfair advantage. After a critical erroneous line call against her in a 2004 US Open quarterfinal match with Jennifer Capriati, she complained to the umpire. After TV replays, it was obvious that the linesperson's call "in" was correct and the overrule by the chair umpire reversing the call to "out" was wrong. This prompted the tennis world to consider introducing the line call assistance technology we now know as "Hawk-Eye." So, it's largely thanks to the egregious call that we have more accuracy in the sport.

Carlos said something placating but stopped short of an apology, which in my opinion was appropriate, and the boos erupted again, getting louder with each serve. At that moment, Serena just couldn't return to that calm mental place of focus that players need to maintain in order to thrive on the court. That was the place Naomi sought to protect on her own side of the court by turning her back and blocking out the unusual distractions. She did not allow what was happening between Carlos and Serena to enter her space.

Serena approached the chair again, calling him a liar and a thief, accusing him of stealing a point from her, resulting in a third code violation—a game. Now the score was 5-3 for Naomi. Neither player seemed to hear that announcement, which added to the confusion, so Carlos summoned them over to the umpire chair. Then the US Open referee, Brian Earley, and Donna Kelso, the WTA supervisor, were summoned to the court. It was right at that moment that I was walking up courtside to take my seat next to the other US Open staff, including the CEO, Gordon Smith, and chief of professional tennis, Stacey Allaster.

Apart from the commentators in the control booths and the millions of folks watching this play out on their television screens, few of us attending the match itself were able to hear the full exchanges or the content of the outbursts. Those us of standing closest to the action were struggling to understand the story that was unfolding, but body language was giving us some clues. Enough so that, as Serena locked eyes with me and walked toward the courtside seats, all I could think was, "Oh shit . . ." I quickly looked away, carefully avoiding eye contact.

Eventually the match continued, Serena held serve and then Naomi handily served out the final game. But it wasn't the mood you'd expect when a young player wins her maiden Grand Slam tournament final, especially against someone she'd grown up admiring. Most of us would throw up our rackets, fist pump our camp, or drop to the ground to kiss the court in pure elation, relief, and gratitude. Not Naomi, who with a stoic expression made a deep, respectful bow to Serena at the net, according to the Japanese custom, then walked over for a handshake.

Clearly, this poor girl thought the roars and jeers from the stands were directed at her. Naomi was raised with a sense of honor and humility consistent with her heritage, so winning on these terms could not have felt good to her. Two great women players of color had, in different ways, suffered in what should have been a great moment in the history of tennis. With people in the stands still vocalizing their extreme displeasure over what had gone down in the final minutes of the match, Serena took Naomi into her arms to give her a comforting embrace, then told the crowd to knock it off. It was a beautiful act of sportsmanship, but it didn't change the fact that Naomi had been robbed of her moment.

As all this was taking place and the USTA officials, myself included, stepped onto the podium to greet our champions, it reminded me of the previous year, when I was on the stage with Madison Keys and Sloane Stephens for the women's final. It was a historic moment to have that many people of color front and center at a finals trophy ceremony, representing both the athletes and the corporate side of sports. I was presenting alongside Thasunda Brown Duckett, CEO of Chase Consumer Banking at JPMorgan Chase. But there was so much going on at present that I don't think anyone realized the significance, apart from myself. Although Venus and Serena had been in many finals on the dais, there had never been other black people who were executives on the same stage.

The boos didn't stop as Naomi made her way toward us. As she stood there flanked by USTA CEO and executive director Gordon Smith to her right and Serena to her left, Naomi pulled her black visor down over her tear-streaked face and lowered her head, still taking the crowd's anger and disappointment to heart. Seeing this, Serena sidled up next to her to give her another encouraging squeeze, then, full of emotion, told the fans, "I just want to tell you guys, she played well, and this is her first Grand Slam tournament. Let's make this the best moment we can."

The jeers were beginning to taper off, but they could still be heard, so Serena continued.

"I know you guys were here rooting, and I was rooting too. But let's make this the best moment we can, and we'll get through it. Let's give everyone the credit where credit's due. Let's not boo anymore. We're gonna get through this, and let's be positive. So, congratulations, Naomi! No more booing."

There was a lull of about twenty seconds, then cheers from the stands. Serena's act of kindness and sisterhood turned things around.

Then it was time for me to speak, and after hearing Serena speak from the heart, I gave the words I'd prepared a quick tweak to acknowledge the mood in the stadium.

"It's not the outcome we were looking for . . . ," I preambled, referring to the circumstances of the penalties and the booing, before presenting the trophies with heartfelt hugs for both players. It was the usual drill when there's a winner standing on the dais . . . but not.

Then Naomi spoke.

"I know that everyone was cheering for her. I'm sorry it had to end like this."

I was moved by her humility and amazed by the fact that she would think to apologize. I was also mortified that Naomi felt that she needed to apologize. For what? Winning? For doing her job and accomplishing her goal in beating her idol . . . again? The tearful looks of empathy on the faces of her family members and coach at the time, Sascha Bajin, said it all.

"I just want to say thank you for watching the match. It was always my dream to play Serena in the US Open Finals. So, I'm really glad that I was able to do that," she added, turning toward her erstwhile opponent. "And I'm really grateful I was able to play with you."

After a series of photographs, the players made their way back into the locker room, to ready themselves for the press gang that was waiting for them in the interview room. As was customary, I headed back to the President's Suite to thank my guests for coming and wish them safe travels. At the end of a long and fraught day I was looking forward to getting rested and recharged for the men's championship finals the following day. Immediately my phone started pinging nonstop. I was trending on social media. Within minutes of my speech people were eviscerating me for my words, suggesting that I'd somehow wanted Serena to be the winner of the match and was not supportive of Naomi or happy for her win. It was the most brutal Twitterstorm of my career, and it of course wasn't even true!

In the US Open official transportation on the ride back to the hotel, I replayed the scene in my head. If I could do it all over again, would I have chosen my words more carefully? Absolutely. But I knew what was in my heart, the social media trolls be damned. What bothered me more than anything was the thought that, even for a moment, Naomi might think I could have been anything less than thrilled for her. I was full of pride for everything she had done that day—not just her prowess on the court but also the grace, poise, humility, and maturity she'd demonstrated during and after the match.

That night, I couldn't sleep. Unexpectedly, I had become part of the controversy involving the match. The thought that anyone could twist my words and think I had an agenda was painful and disappointed me. By midmorning the next day, news pundits were revving into full gear regarding the previous day's match between Serena and Naomi. I was even summoned to be on the ESPN set with Chris McKendry and Mary Joe Fernández, where we clarified my remarks, to avoid ill intent. The media frenzy all but eclipsing the men's tennis and that was a shame, because Novak Djokovic made a powerful comeback against Juan Martín del Potro. People were almost as divided over what happened in the women's final as they were about the last US presidential election. Families were fighting over who was right and who was wrong. Editorials were being written and people around the world were taking their positions and turning this into a standoff between the politically correct and the anti-feminists.

If you took Carlos Ramos's side, apparently you were a racist misogynist. If you defended Serena, you were somehow knocking Naomi's victory and dealing a blow to sportsmanship while enabling a spoiled, entitled brat. Neither characterization of what happened was true. One Australian paper even ran a deeply offensive cartoon of Serena stomping on her racket that echoed back to the Jim Crow–era caricatures of black people. It was sickening. The whole media maelstrom was a race to the bottom. "Serena-gate" had somehow become a flash point for all the political divisiveness that has been tearing this country, and the rest of the world, apart.

I empathized with all the players in this drama, including Carlos, whom I've always found to be professional, thoughtful, and measured. While we haven't spent a lot of time together, Carlos clearly cares for the well-being of the players as much as for the integrity of the game, even though I didn't entirely agree with how he handled this particular match.

At the end of the trophy ceremony, we decided to not present Carlos's token of appreciation on the court as was customary. It was no reflection on what we thought of his performance. With the crowd's energy that night, we decided it would be best to recognize his service in the back of the house. We wanted the attention to be on the athletes, specifically on our champion, Naomi.

I understood how the situation may have looked from his vantage point. Patrick Mouratoglou, Serena's coach, was indeed using hand signals, which he admitted to later, so Carlos was simply doing his job. But Serena said that she didn't see what Carlos saw as coaching and got upset at the implication she was cheating, bringing forth all the frustration from unfair judgments of the past into that moment on center court. Perhaps if Carlos had given a soft warning about the activity in both boxes from the start, the later escalation could have been avoided. We'll never know.

Serena's strength, unique style, and outspokenness had often made her a moving target. Without a doubt, there were occasions when she'd been at the receiving end of unconscious, and not so unconscious, bias. All the grievances of the past twenty years of her professional career erupted from that place. But after Serena called Carlos a "thief," attacking the integrity of the chair umpire, she left him no other choice but to implement the Grand Slam rule with a code. Had it been the first offense, it would have just been a warning. But it wasn't; it was the third violation, and in accordance with the rules, it warranted a game.

If any good came out of this situation, it was an increased awareness among members of all the major tennis organizations: the USTA, the International Tennis Federation (ITF), which sets the rules and oversees the sport globally, the Association of Tennis Professionals (ATP), which represents the male players, and the Women's Tennis Association (WTA) that rules need to be implemented more consistently from tour to tour. Officials at each of these organizations have different relationships with different players. Soft warnings may be more prevalent among some umpires than others, for example.

But by far the most powerful lesson was one of sportsmanship from the athletes themselves, which moved me to weigh in with my own statement:

> What Serena did on the podium today showed a great deal of class and sportsmanship. This was Naomi's moment, and Serena wanted her to be able to enjoy it. That was a class move from a true champion. What Serena has accomplished this year in playing her way back on to the tour is truly amazing. She continues to inspire because she continues to strive to be the best. She owns virtually every page of the record book, but she's never been one to rest on her laurels.
>
> She's always working to improve; always eager to embrace new challenges, and to set new standards. She is an inspiration to me, personally, and a credit to our sport, win or lose. I know that she was frustrated about the way the

match played out, but the way she stepped up after the final and gave full credit to Naomi for a match well-played speaks volumes about who she is.

I meant every word. Did she go too far in her passion? In my opinion, yes. While her feeling was valid that umpires often make calls against female players and I am glad it sparked a discussion about equity in the game, Serena had lost control. Arthur Ashe Stadium, in the middle of the final, was the wrong place to make the right point. Had she not allowed her anger to get the better of her, had Serena's mind stayed focused on her playing, she might have made the most spectacular comeback in the history of women's tennis.

I also think Carlos could have done more to diffuse the situation. He went strictly by the book, but there were opportunities to acknowledge what Serena was feeling and smooth things over, one human being to another. Everyone was too entrenched in their positions, allowing their defensiveness and emotional baggage to distract from what mattered most—the match itself.

Again, it was not my role as head of the USTA to pick a side, although plenty of people were trying to draw me in following that ESPN interview. Being African American, many people assumed that I was favoring Serena. But how could I when both women were black? I decided the best way forward was to appear on a national news show I knew to be tough but fair, and so I accepted an invitation to appear on *CBS This Morning* on the Tuesday following the finals. Hoping to bring some much-needed clarity and perspective on what happened, I put on my best game face. That morning, I had a lot of interviews on my schedule, including a taping of *We Need to Talk* on CBS Sports Network, where I am a cohost. But first, I had to walk through the ring of fire, so I got myself up extra early, put on a bright-red dress with matching scarlet lipstick, and headed to the CBS News studio in Manhattan. Despite my nerves, I was looking forward to seeing Gayle King, who I knew from attending various events and galas over the years. As a black woman, she, more than anyone in media, could relate to my situation and understand the world I was moving in.

Sitting at the table with Gayle, Norah O'Donnell, and John Dickerson, I explained how Carlos was simply following the code. I made it clear that, in my opinion, he could have done things a little differently, starting with a soft warning for the coaching, which I also pointed out was prevalent among both men and women players. He was feeling defensive and fed up. He was being human, as was Serena. I explained that there needs to be more consistency in the way rules are interpreted and applied, particularly for male and female players. I came right out and said I don't believe women are always treated fairly. While I took care not to undermine Carlos, I was transparent about my belief that, in general in our sport, unconscious bias can influence the way matches are officiated.

Gayle, John, and Norah were persistent. They'd done their homework, wanting me to come out and say I thought what had happened at the women's final was unfair. They asked me to explain, for example, how Serena could have been fined $10,000 for verbal abuse when, in 2009, Roger Federer was only fined $1,500 for language that was arguably much worse. (A hot mic had picked up some choice obscenities in an exchange with the umpire during the US Open Final.) But also, Serena's was a third offense and Roger's was a first. Each offense has an escalating fine.

I had to clarify some common misunderstandings about the fines, which were at the discretion of the US Open Grand Slam tournament administrator, *not* the USTA, based on the warning, the point, and the game, and that the fine could be appealed. In our role as host and facilitator of the US Open, it's our job to make sure that the fans, players, umpires, and ball persons are safe and well cared for. As the national governing body for tennis in the United States—a nonprofit organization, the USTA's main role is to promote and develop the growth of tennis at the grassroots community level on up. Part of that includes the running of the USTA Billie Jean King National Tennis Center, where the US Open takes place. We oversee operations, establish how we want to present the event, and we cooperate with the other Grand Slams—Australian Open, Roland-Garros, and Wimbledon—as it relates to the rules and code of conduct of the competition.

At the heart of the tournament is the integrity of the competition, and this is where we are hands off when it comes to the matches themselves. As far as the public is concerned, we may be the face of the US Open, but we do not make line calls or take any other actions that determine who are the winners and losers. Officiating is managed by the US Open team and the competition is overseen by the tournament referee in collaboration with the Grand Slam supervisor. But, when called to the court, the tournament referee and the Grand Slam supervisor cannot overturn a chair umpire's decision as long as their decisions are implemented in accordance with the Grand Slam Rule Book. For this match, when the tournament referee was called to the court, it was determined that Carlos had called it in accordance with the rules, and his decision was final.

Of course, there was a little pushback on this from the CBS team. Comparisons were made with another umpire doing the exact opposite of Carlos Ramos and jumping off his chair to offer support to Nick Kyrgios, going against US Open protocol. They had their gotcha question ready:

Was that not also an example of the USTA rules being unevenly applied?

"[That umpire] was penalized, he was docked, you never saw him back on a main court," I countered, adding, "These were global chair umpires who worked for the ATP World Tour, ITF, USTA, and WTA Tours. It's not about the USTA. We have the best umpires in the world who come and work for us at every Grand Slam."

It's hard for even the most die-hard tennis fans to understand which organization is responsible for various aspects of an event. It was my job as the president of the USTA to help the public understand what we do as providers of the venue for the US Open, which has many moving parts and involves collaboration with multiple agencies, both national and global. I drew from my training as a player and a commentator while I was on the hot seat, smiling, and patiently explaining the complexities that neither the CBS News team nor the general public could be expected to know.

The press attention had been relentless and was showing no signs of slowing down. I'd played back the events of the previous few days enough times in my head, questioning myself and wondering how things could have been handled differently. But I was grateful to have the opportunity to clarify a few points. I hated the idea that these last finals on my watch could have somehow clouded the reputation of the USTA and my legacy of fairness and inclusion during my four years at the helm.

But I was especially glad to be able to talk about Naomi. At the start of the interview, I tried to steer the conversation around to her.

"Naomi was the champion in all of this, and her moment was overshadowed," I told the morning show crew. "It's important that we celebrate her."

Everything I'd done leading up to this point as chair and president of the USTA had been about nurturing young talent and giving young players like Naomi the chance to shine. The only playback that I was interested in at this point was Naomi's brilliant performance on the court.

There had been moments throughout my career as a player when I believed I was being overlooked, so I understood how lousy she must have felt.

When I was sixteen, the college scouts started reaching out to everyone but me. Each year, the recruiters came to the Girls' 18-and-Under USTA National Indoor tournament to check out the players. As one of the top juniors in the country who had defended her Illinois high school state championship title as a senior a month earlier, I was looking forward to having a few conversations. I had my eye on UCLA or USC, top-ranked colleges where I could play outdoors year-round in the California sunshine, although Northwestern University was always top of mind. But every coach there was acting like I didn't exist. They were approaching many other players whose stats were nowhere near as impressive as mine. The only one who appeared to acknowledge my existence was Sandy Stap Clifton, the Northwestern University coach who was already in the process of recruiting me. Each time I won a match, she'd smile and quietly congratulate me.

Being ignored went on for three days. It was torture.

Not being assertive enough at that age to ask questions, I felt ostracized. I even wondered if it was racism, because I was the only African American in the tournament. I wasn't usually inclined to jump to a conclusion like that, but being unfamiliar with how the process worked, I couldn't think of any other reason. I didn't realize it was because the scouts assumed that at age sixteen I was not a high school senior (the internet had not yet been invented). But Sandy knew. That's why she kept winking at me. I was well known in the Chicago area, but the other recruiters couldn't have known. Sandy didn't want to tip them off by making a fuss of me because she wanted me for the NU Wildcats.

Once the scouts realized I was indeed of recruitment age, the floodgates opened. I'd already made my commitment to Northwestern by then. Besides, I was still a bit offended. Whenever I was playing against USC or UCLA, I must admit I hit the ball just a little bit harder because I never forgot that feeling of being left out in the cold.

I also knew how it felt to hear thousands of rowdy New York tennis fans boo in your general direction. My first appearance on center court as USTA head, I stepped out behind Mayor Bill de Blasio. It was protocol that as leader of the city he would walk out first, as many mayors had done before him, for the opening ceremony. But the jeers were so loud as he tried to welcome the crowd to the 2015 US Open, you could barely hear him.

New Yorkers will always be New Yorkers—brash, opinionated, and extremely vocal about their beliefs. Add to their intrinsic nature several hours of drinking under the hot sun, their emotions whipped up into a frenzy as they watch their favorite players, and you end up with a volatile cocktail of audience reaction. These emotional eruptions from the crowd were by no means unfamiliar to me. Yet I wasn't quite expecting that reception. Although I quickly realized the boos weren't intended for me, appearing in front of the stands in full makeup and black-tie gala dress for my USTA debut was meant to be my big moment. This was nothing like how it played out in my head while I was getting my hair done. When you're at the physical center of a full blast of negative energy, it's truly unnerving.

So, I had nothing but empathy for Naomi. I was relieved when Gayle finally shifted the conversation back to who mattered most.

"My heart broke for her feeling that in that moment she felt like she had to apologize even for winning," she said. "Everyone, even you Katrina at the beginning, sounded like you were saying you were sorry things ended this way meaning that Serena didn't win. Here is this young woman standing there during the best moment of her career and it seemed like everyone in the building was against her at that time. She couldn't even have the celebratory moment when she won."

Gayle set up the shot for me.

"My words were misconstrued," I responded. "When I was talking about the outcome, I was referring to the behavior of the fans, not about who won. But this is the biggest moment of your life, you are a US Open champion, you are a Grand Slam tournament champion, for the first time. You are the best player, and, because of the turn of events, the crowd is booing."

I hoped that Naomi was out there somewhere, listening. How she must have been feeling about all the toxicity was haunting me. I needed her to know that all the chatter had nothing to do with how much I admired her. I fired off a text to Sascha, Naomi's coach, whom I'd known since his first days on the tour as Serena's hitting partner:

Sascha,

Congrats to you, Naomi and your entire team. What an accomplishment 😊! Naomi played amazing the entire tournament and got better with each match. It's so unfortunate how things ended, the interferences and behaviors, and that is what I was referring to when I mentioned the "outcome" of the match. Not referencing the result. You know I have mad respect for you and ALL players. Unfortunately for all of us, the ending was weird, the commotion was disruptive, and Naomi didn't get a chance to truly celebrate and receive the love that she so rightly deserved. It's been a huge debacle, but I want Naomi to know that she is a true champion. She was my dark horse before the tournament, and she IS the future. She will earn many more titles in the future. Congrats again. You both deserve to be applauded . . .

As a former player, I had many special relationships with the players and coaches. This wasn't about my role as representative of the USTA. This was strictly between me and her. Knowing Sascha, I expected to hear back immediately. I did not.

After firing off that text to Sascha, the following day, I woke up to crickets. Radio silence was not like him. Perhaps he and Naomi were distracted by the flurry of press interviews and events that follow winning a championship, but . . . still. I couldn't help but worry that the non-response was because they were upset with me.

After a week caught in the viral echo chamber of social media, I couldn't take it anymore. I'd already suffered a week of sleepless nights. I was in Croatia, for the Davis Cup semifinal tie between the USA and Croatia, but my head was still at the US Open, ruminating on what had gone down. I sent Sascha another text asking him to confirm whether he'd received my text, to which he immediately replied:

Hey, I'm so sorry I didn't respond, but I received like 300 messages after that final and yours was lost in it ☹. I hope you understand what I mean. Thank you so much for your kind words. It truly means a lot to me and I will tell Naomi you reached out to me twice now to make sure she gets your message.

To which I replied:

> You're a rock star and I totally understand. I was just hoping you weren't mad at me knowing how much I love you and Naomi. It was a whirlwind after the final and my words were misinterpreted. I have felt horrible ever since and just wanted Naomi to know directly from me how much I adore her and NOT via social media, which can be so cruel.

Sascha sent me Naomi's number and we finally connected. Everyone but the most important person in this story took my words and intentions and twisted them somehow. But Naomi already knew what was in my heart. In an exchange of texts, she told me it had never even occurred to her to take my remarks at the US Open Women's Final any way other than as intended, and that she was thrilled to hear from me directly.

The most timeless lesson from what had happened that day was far from lost on the two people who mattered most: Serena and Naomi. A profile of Serena in the August 2019 edition of *Harper's Bazaar* gave me chills. The article reprinted an exchange of text messages between the two athletes that demonstrated a compassion and respect for one another that's the true definition of sportsmanship:

> Hey, Naomi! It's Serena Williams. As I said on the court, I am so proud of you and I am truly sorry. I thought I was doing the right thing in sticking up for myself. But I had no idea the media would pit us against each other. I would

love the chance to live that moment over again. I am, was, and will always be happy for you and supportive of you. I would never, ever want the light to shine away from another female, specifically another black female athlete. I can't wait for your future and, believe me, I will always be watching as a big fan! I wish you only success today and in the future. Once again, I am so proud of you. All my love and your fan, Serena.

Naomi replied graciously. Her response touched Serena so much that tears rolled down her cheeks. The one thing about that match that haunted her most was how Naomi had lost her moment. But the younger athlete was finally able to give Serena peace.

2

ALL EYES ON ME

Spectacular performances are preceded
by spectacular preparation.

—FRANK GIAMPAOLO, TENNIS COACH

There are occasions when you walk onto a tennis court and it feels humongous and the ball appears as large as a grapefruit and you couldn't miss if you tried. There are other times when the opposite happens and the court feels incredibly small and you can't keep a ball inside its lines no matter how hard you try. I sometimes had these opposing experiences when I approached my job as president of the United States Tennis Association. The task before me can seem enormous and at other times, manageable.

One thing I was clear about as president was that I was not placed in the position because of my differences, as a black woman. If anything, I was there in spite of them. I believe that my ascension to the USTA President was a testament to my ability to bridge the gap that my differences so easily create—particularly in a culture that for so long did not exactly embrace being different.

I believe that uniting around a common purpose can successfully drive us forward, while simultaneously strengthening the bonds among those who are working to achieve that success. This should be the approach in all organizations and corporations. But in order to truly succeed, you have to be honest with who you are and stop accepting what others want you to be.

I had a lot of pressure being the "first." Being the first former player, the first African American, and the youngest. All eyes were on me, as I would likely be scrutinized more because it was an unfamiliar territory for those around me. I had to be extra knowledgeable of decisions and how I represented the organization, especially being black. We have always had to work twice as hard to get recognition for our accomplishments.

An example was putting a roof over Arthur Ashe Stadium, which had been debated by USTA board after board for nearly a decade. It was scheduled to be finished during my tenure, and I wanted to make sure it was completed on budget and on time. Global warming had caused the weather patterns to change, presenting more rain delays and inconveniencing players and fans. For five consecutive years, we ended up with a Monday final—three unplanned and two planned owing to television contracts. Obviously, there was a business cost to this, including ticket refunds, overtime, and other added operating costs. Not having a retractable roof also put us behind other Grand Slam tournament venues, like Wimbledon and the Australian Open in Melbourne. It was a matter of pride for us to have the best and most cutting-edge facilities. But covering a stadium of that vast size—it's built on a landfill—would require a creative solution, because a standard design would be too heavy. We went through countless architectural rendering until we finalized a design with Rossetti, an architectural firm based in Detroit, which was the architectural firm used for the original stadium, twenty years earlier.

The work of construction began under my predecessor as president of the USTA, Dave Haggerty, who oversaw the foundational work, including all the pilings that were installed underground to support the structure. By the 2015 US Open, thirteen million pounds of steel had been erected. By the following year's tournament, we had a fully functioning, retractable roof.

It just so happened that in our first year with the roof we experienced some of the hottest temperatures—99 degrees day after day—creating a sauna effect for players and people in the stands, who weren't getting the full breeze from outside. We had a coolant system in the form of cold water running through pipes to bring the temperature down, but no air-conditioning. However, everyone was able to enjoy continuous play, with no interrupted television broadcasts.

It was a Tuesday night, during a first-round singles match between Spaniard Rafael Nadal and Italian Andreas Seppi (which Rafa won), when we first rolled out the $150 million roof. I'd left, because I had an early meeting the next morning, but I listened to the match on ESPN radio as I was riding back to my hotel, then pulled up ESPN3 on my phone because I needed to see this. I could hear the crowd cheering as that giant structure sealed out the rain in all of five minutes and twenty-two seconds.

Getting that roof in place was the culmination of years of preparation and collaboration at its best. Even though it's been cited as one of my major accomplishments during my tenure at the USTA, I don't see it that way. As a black woman, I have never regarded leadership as a singular pursuit. It's always been about enabling others to be their best and do their jobs. That mindset has been affirmed by tennis, where we are not organically playing against each other so much as with each other, particularly in doubles but also singles. It's a singular ball that the players are dancing with, in confined space. Whatever side of the net we're on, as players we are all working toward the same goal of playing a brilliant match.

That sense of working together toward something greater than ourselves was amplified when I became the leader of an organization. My role was not to take credit or control but to collaborate and give people all the support they need to be successful. Although in many ways I was the face of the USTA during my tenure, I was always acutely aware of the importance of the people making it happen behind the scenes, from budgeting and planning for junior and adult programs, leagues and tennis tournaments all across the country, to the logistics of putting on a Grand Slam tournament on the global stage of New York City.

What most tennis fans don't necessarily know is that so much intricate planning takes place before that final moment when champions are awarded their trophies on the dais. A solid six months of work goes into that two weeks at the end of the summer. As Gordon Smith, the USTA's CEO and longest-serving executive director puts it, "We put on the Super Bowl every day for fourteen days." Well, more like twenty-one days if you include Fan Week that takes place the week before, which is busy too, at no cost to fans.

The window is short, some might even say nonexistent, between the end of one Open and the start of the other. The planning for the next one begins before the current one is over. The week after the Open has ended we're on conference calls and in meetings dissecting what went right, what was missed, and where unnecessary resources were wasted and what can be done better, whether that's providing enough buses and cars to transport our players, their entourages, and our VIP guests, having enough security detail at the entrances, or training enough seasonal workers to greet and guide people as they enter the grounds. By November, the budget is in place and by May we are already creating menus, tweaking our interior design, selecting artwork for the suites and grounds, and putting together our invitation list. In short, it's a lot of work. I let my delegating skills shine, and with teamwork we cast a beautiful glow on all our efforts.

The US Open is the USTA's giant bake sale, so it's essential to put our best face forward to members of the general public as well as to our VIPs. Meetings and entertaining international and local guests in the President's Suite are a big part of my job duties during the final days of the US Open, when our delegates arrive from all over the world, along with executives from the ITF, the WTA, and the ATP, for critical conversations and year-end honors. The Open is when we recognize and thank our sponsors and all the volunteers who work tirelessly, year-round, promoting the sport and supporting up-and-coming young players throughout the country.

Creating a welcoming environment that makes guests and delegates from around the world feel completely at home is why the President's Suite exists. But by the time I stepped into my role at the USTA, it had been decades since some of the areas had been redecorated. The carpets were shopworn and the place had a dated, 1990s clubhouse look. We hired New York interior designer Everick Brown to give us a face-lift.

Our intention was multifaceted. We wanted to freshen up the look for our more progressive membership while not alienating the old guard. We wanted to be inclusive yet respectful. Instead of classic traditional, we opted for a modern classic environment with clean lines and neutral tones to showcase revolving sections of artwork—pieces by women and African Americans and artists from around the world. It reflected some of who I am and where I come from, while including other cultures.

The main changes we made were to the seating and layout. The previous space felt more like a lounge where guests would sink into a comfy chair with a cocktail in hand and hardly move for most of the afternoon. Everick built a whole new set of furniture, intentionally creating sofas and chairs that would not be quite as inviting for guests to perch on for an entire afternoon, although my dad did find his own little corner. We also upgraded the bar, with modern wood finishes. The intention was to have our guests get up, circulate, and be a more engaged tennis audience. We even removed the barstools to keep the flow, not that all our older members were thrilled about that!

Other hospitality details revolved around food. Getting it right was personal to me. I wanted to re-create that sense of graciousness my parents had taught me early on. Our home was a revolving door of friends, family, kids from the neighborhood, or basically anyone who touched our lives and could benefit from warmth, friendship, and one of my mother's legendary Sunday feasts. I've lost count of the hundreds who've passed through our kitchen for extra helpings of collard greens, sweet potato pie, or macaroni and cheese. Anyone from my tennis family who happened to be passing through Chicago had a standing invitation. Gladys Knight became fast friends with my parents and even invited herself over for dinner.

My parents took that hospitality with them wherever they went. When they came to watch me play while I was on the professional tour, Mom would find a way to serve up her Sunday best. At Wimbledon, we used to rent houses or apartments near the courts. Mom would then scope out the nearby West Indian supermarkets, which often had the ingredients she needed for her soul food recipes. Since most of the players stayed within the same few blocks, word quickly spread that Mom was cooking. Mostly all the players of color would come, including Venus, Serena, and their mother, who didn't love English food. The tiny dining rooms of these rentals were too narrow to contain everyone, so guests piled food on their plates, then filled up every room in the house.

These intimate feasts were partly about my bringing my culture, where it is important to extend an offering of food and present a welcoming environment. It helped me to create a bond with players on the tour and to establish a fondness for teamwork. It also made me feel less alone. This had been a tradition by the players of color long before I played at Wimbledon.

Food is a way to make everyone feel at home. With a more diverse group of guests in the President's Suite, we wanted to make sure everyone could find something they loved, whether they were Roland-Garros dignitaries craving escargots or Julia Steele, the queen of Chicago Prairie Tennis Club, the oldest African American organized club. Miss Steele was never one for fancy food and felt uncomfortable putting something on her plate that she couldn't easily identify, so she was overjoyed to find collard greens on the menu.

As luck would have it, on a tournament day when Miss Steele was in the house, the theme was "Southern Cuisine," prepared by guest chef, Carla Hall, working alongside executive chef, Jennifer Cox of Levy Restaurants, the President's Suite head chef.

"Oh my goodness! Greens and fried chicken!" she exclaimed when she saw what was on our chef's luncheon table. "That's it. I've arrived. Y'all just made me feel so welcome!"

That was exactly the kind of reaction we were going for on any given day or night.

It all serves a larger purpose. People only see us, the USTA leadership team, sitting with stars enjoying the best seats in the house. Sports channels and news cameras pan to celebrities like Mariska Hargitay and Lin-Manuel Miranda looking enthralled by a match as they sip cocktails. What they don't realize is that figuring out who gets invited to the President's Suite and where they get to sit in the box is highly strategic, and it's not just about the rich and famous. This is one of our main vehicles for fund-raising to benefit tennis organizations around the country. Yes, the high-profile tennis fans bring more eyeballs to the Open, but it's mostly about honoring and thanking the sponsors, donors, and volunteers of local Community Tennis Association programs or National Junior Tennis and Learning (NJTL) network chapters who further our mission. In order to ensure that everyone comes away feeling acknowledged and enriched by the experience, we pay attention to the smallest detail, with nothing taken lightly.

The USTA has hosted more than five thousand people a year in the suite. Everyone in the world wants a piece of the US Open, including heads of state, Wall Street bankers, NASCAR drivers, NBA players, Washington politicos, actors, singers, and business tycoons. It's probably the most celebrity-centric event in New York, which is saying a lot.

Beyond the President's Suite, we have the entire USTA Billie Jean King National Tennis Center to prepare. Having just come off the overhaul of the Flushing, Queens, site, there was plenty that could have gone wrong. We had to think in new ways about ventilation, for example, and the fact that covering the courts without giving the indoor atmosphere time to cool down could make conditions worse for the players.

The introduction of new features, like the serve and warm-up clocks, which speed up the pace of play and ensure a consistent set of enforcement standards, also took some familiarization. The 2018 US Open would be the first Grand Slam tournament in the world to adopt the technology, but we knew it would be controversial. We weren't sure, for example, how it would be accepted by Novak Djokovic or Rafael Nadal, who like to take their time between playing points. There has always been a twenty-second rule between points at the majors, but it hasn't always been implemented equally. This particular year, the rule was changed to twenty-five seconds so the players could get adjusted better.

The USTA has always bent over backward to look after its Grand Slam tournament players, beyond just their experiences on the court. It provides gourmet meals, transportation, as well as security on site and at off-site functions, where necessary. This VIP treatment isn't limited to the biggest stars. It's for every professional player scheduled for matches at this event. We've all benefited from the hospitality of these big tournaments. The difference now is that even the rookies can walk away with a nice check, because today the purse is $50,000 for losing in the first round, compared to $15,000 when I was playing professionally.

Separately, as a person of color, I am proud that we provide more resources and exposure to the grassroots players from underserved communities whose talent needs to be nurtured with top coaches, more court time, better equipment, and better access in general through our National Junior Tennis and Learning chapters. I would love to see more kids, like the hundreds I've coached and mentored over the years, get to the level of a Grand Slam tournament, or any professional tournament, and experience what it's like to be at the center of the tennis universe, where they are served by teams of experts whose sole focus is their well-being and comfort.

We consider every detail, from how many heat breaks during a match to the number of towels to supply to each player (about four per player unless it's Rafa, who goes through at least a dozen). We even think about how many times to prewash the towels to lose that slick, new-towel texture and allow for maximum absorption. Imagine what this kind of first-class care and deference could do for a struggling player's self-esteem? This is what many players who qualify for the first time experience. Whether it's cooling stations, ice vests, hydration, or restroom breaks, our player services team relies on professional medical opinions to help us anticipate exactly what the players' needs will be minute by minute.

Then there are the logistics of transforming what is one of the busiest public tennis centers in the tristate area eleven months of the year into a world-class professional facility with the fine dining and luxury retail and guest services you would expect as one of the eight-hundred-thousand-plus ticket holders who pay up to $800 a pop, face value, based on the day of the event, often flying in from across the country and the world to enjoy the best that tennis has to offer.

All this work gets done down to the wire. People don't see the scrambling that goes on behind the scenes; they have no idea we're holding our breath seconds before the gates open. When local tennis players walk into the location in early August and see what looks like a construction site, they cannot imagine everything being finalized and ready just weeks later, with all the flower arrangements in place, greeters trained, and original artwork hung by showtime. We go through a lot of sleepless nights in the weeks before, because no amount of planning and precision can prevent last-minute hiccups. We have to build in a certain amount of flexibility and prepare to pivot when nature, or human nature, interferes with those carefully laid plans.

It's why I never make the assumption I've got this and always seek out the expertise of others. In fact, I walk through life with my own personal board of directors: friends and colleagues with certain skills who I can lean on and not be shy about asking certain questions of. Friends who are experts in business, finance, marketing, law, and so on. Individuals who I can lean on in certain disciplines to help me to understand what I don't know. They give me the confidence to delegate well because we can't all be good at the same things. They also help me to break through that clutter and give me clarity when I need it most. They are in my corner, supporting me with unconditional loyalty and experience that's different from my own.

These personal board members never "yes" me. Rather, they challenge me, often playing devil's advocate. They'll tell me the truth in a supportive way. If they disagree with something, they'll say something like "You might want to rethink that one." Sometimes they will simply act as my sounding board, allowing me to crystallize my thoughts on a subject. The conversations are productive, never negative, giving me a high level of expertise. They also show up for me—always. I don't call on everyone regularly, so they know that when I do reach out, it's for something important.

Being involved at this level of sport is a privilege. I never could have imagined it given where I came from. I never would have made it this far without the support, teamwork, and encouragement of others.

Most mainstream tennis fans would not have heard of Bunny Williams, but she's one of the former women supervisors and former gold badge chair umpires (highest status), along with Woodie Walker, I admire most. After working for the USTA for more than forty years, Bunny retired at seventy in 2018; Woodie retired in 2017, well into her eighties, as the US Open chief umpire. Both women were such maternal figures on the tours, as well as great leaders. Early on in my career, when I was playing the ITF pro circuit, I often encountered Bunny and Woodie on the road. They always made sure I was taken care of at events. I'm not sure if it was because they found me charming or because I was a minority in places they knew could be challenging. I never asked the ladies, although I always let them know how much I appreciated their care and concern.

A few years later, when I made it onto the WTA Tour, Bunny and Woodie, whose umpire careers were evolving and taking them to larger events, were also there. In fact, Bunny was the umpire at a quarterfinals singles match I was playing in February 1994 at the IGA Tennis Classic in Oklahoma City against my doubles partner, African American Zina Garrison.

Five years later, Bunny was head of one of our biggest American tournaments—the Family Circle Cup in Hilton Head, South Carolina, where she resides. Bunny had me over to her home, where she greeted me like an old friend, full of graciousness, warmth, and Southern belle hospitality.

People like Bunny looking out for me from the juniors to the professional level helped a great deal in not making me feel like an outsider. At times this feeling was softened by tennis being an international sport, and there being a lot of other "only ones." But the warm embrace of individuals from the tennis establishment went a long way toward that feeling of inclusion.

My career journey, in the world of tennis and the higher echelons of sport often put me in rooms where I was the only woman, only black woman, or only person of color. I have gained an incredible amount of knowledge being in these rooms. I have also learned that it is especially important to bring my full, true self into the conversation—culture and all. But it took many rich and varied experiences for me to recognize this truth.

I started out in the park. As a junior player I wasn't one to emote outside tennis. Our house was full of laughter, but my parents were always even-tempered and frowned upon extreme displays of emotion or "silliness" as my dad called it, so in daily life I learned to internalize. Although my mom was full of personality and missed her calling to be an actress, my dad didn't like our "joking" around too much. Even though he was a funny character himself. But it would all come out on the court. Whether I was frustrated or elated with joy, you'd know it while I was in the middle of a match. As a junior player I would often lose my temper when I missed a shot.

Before I could get to the next level, I had to master my emotions. Winning helps. But I developed self-control by establishing little rituals between playing points, like taking a deep breath, straightening the strings of my racket, and visualizing what I needed to do as I stepped up to the service line. As I moved forward, bouncing the ball, I'd tell myself, "Hit your target." If I'd netted a ball, I'd remind myself to get down lower or hit up on the ball with more spin next time and see success as the outcome. That calm and focus enabled me to regroup and adapt to the changing circumstances, finding different approaches to challenges—skills that carried over into my careers off the court.

I can thank my earliest mentors for those specific tools of self-leadership: my parents, my first coaches, my teachers, and my pastor. I was raised in the church, which taught me to shake it off and forgive. Mom and Dad would continually remind me that, though I could not control how other people behaved, I had a choice in how I responded to any circumstance. It was never truer than now, with how I responded to a myopic umpire or social media twisting my words. They instilled in me the importance of letting things go. They also showed me that giving others the benefit of the doubt, taking that extra step of walking in someone else's shoes, then moving on, is a demonstration of character.

It's a level of empathy that's empowering. In order to play strategically, for example, you need to appreciate how things look from your opponents' side of the court. It's possible to anticipate their next move when you take the extra step of knowing what type of player they are, understanding their history and where they are coming from. This is also true in business.

Similarly, compassion and forgiveness free up my mind to focus on what's in front of me. It creates a kind of mental clarity, enabling me to look at situations through multiple lenses, which became increasingly important as I scaled the heights of organizational leadership. A big part of my role as head of the USTA was to listen carefully to competing and sometimes contradicting points of view, weighing options as people presented their arguments for a particular course of action.

I kept my balance, holding myself to a higher standard because, as a woman of color, I knew how well it had served me, earning me respect from people of all backgrounds. My even-temperedness and poise beyond my years had real currency, consistently catching the attention of influential people in the tennis world, from women like Bunny, Woodie, and Billie Jean to another key figure on my personal board of advisors: "Uncle" Jim Kelly, chairman of the board of the Harlem Junior Tennis and Education Program (HJTEP).

An African American business owner whose company is a shareholder in Hudson News, the airport bookstores and concessions, Jim has been successful in a number of businesses, from banking to communications.

Uncle Jim, a huge tennis fan, took an interest in my career at an American Tennis Association (ATA) event when I was sixteen. Toward the end of my career, when I was losing more than winning, he sponsored me, keeping me afloat with a combination of my own income and his money. Not only was that investment a demonstration of his faith in me as a player, it was a way of encouraging me to push past the physical pain and disappointments of defeat, instilling in me the resilience to keep moving forward at every stage of my career.

While I was still commentating and coaching, Jim served as a longtime board member of HJTEP alongside Roberta Graves and Bob Holland, former CEO of Ben & Jerry's, under the then chairman, CBS reporter Tony Guida, and others. Roberta recommended that I would be a good choice as the next executive director of the organization in 2005. Jim, who later became the chairman of the board, and other members agreed, and the decision was made—prompting me to move from Florida to New York and giving me my first role as an organizational leader.

Jim and the rest of the board went out on a limb for me, giving me guidance and time to learn how to put together a full budget and manage people. They were instrumental in helping me to find the necessary faith in myself that I could do this. I could bring value as a leader and decision maker. The nonprofit was going through a major transition and was faced with possibly going under, so they needed a fresh perspective from the professional tennis and coaching world.

The HJTEP board members allowed me to supplement what was then a relatively modest nonprofit income by doing other tennis-related assignments. They also encouraged my board positions, understanding that each role helped to elevate the other. I'd always wanted to give back to my sport in larger ways. I wanted to develop skills like fundraising, administration, and people management while leveraging the perspective I'd gained having gone through all levels of the tennis world from grassroots community competing to top-ranked professional.

Alan Schwartz, USTA President 2003–2005, also saw potential in me from the very beginning and has been instrumental in my professional rise. I've known the gentleman I call "the Godfather" since I was a child. Alan was owner of Midtown Athletic Clubs and the initial Midtown Athletic Club in Chicago, one of the largest indoor tennis clubs in the world and one of the first diverse tennis clubs in Chicago's city limits and most welcoming in America. Its membership included a few prominent African American executives and professionals. Having seen me play, Alan offered me free court time during off-peak hours, when there was availability. There weren't many indoor courts available to me in Chicago, and his generosity was a godsend. When I was eight, Alan invited me to join the Chicago District Tennis Association Excellence Program, where he served as a board member.

In 2003, Alan appointed me to the USTA Grievance Committee, which needed representation from an elite athlete. In 1998 the US Olympic Committee, the governing body of all organized sports, mandated that all National Governing Boards (NGB) include a percentage of elite athletes among their directors. It was a controversial decision, but it was an opening to shake up the staid organization and force its members to see things more from a player perspective.

Up to this point, I'd been a player, a coach, and a television commentator. I lived and breathed my sport from the player perspective. But the executive tennis world, where board members with corporate résumés made decisions inside oak-paneled conference rooms that would impact the lives of junior and professional tennis players—folks like me—felt too remote.

As a coach I didn't have the best experience with the USTA, which I viewed as a bureaucratic behemoth that didn't really care about us as athletes. But whatever Alan asked me to do, I did. He convinced me that it was a good opportunity. Alan, along with Billie Jean King, another mentor, convinced me that I had to step into this arena. Sometimes the best way to bring about change is from within.

Joining that first committee was the beginning of my ascension in the USTA. Two years later, Alan nudged me to fill out an application for the USTA Nominating Committee, which is one of the most powerful committees because this is the group that nominates board members. When that process was over, I was invited to join the board instead, as one of three Elite Athletes.

It was intimidating at first, because I didn't always feel like I had the necessary expertise to speak up. Alan, who sat on the USTA board as an immediate past president, sat beside me, giving me a nod of encouragement when he sensed I had something to say. So did former New York City mayor David Dinkins, my "second dad" and one of only two other black people on the board. Martin Blackman, being one of the other two Elite Athlete representatives, as a former player. David taught me all the intricacies of navigating that room.

I learned from him to read all the materials, take notes, and be prepared. I also learned to not be afraid to speak up.

One of the best pieces of advice he gave me was that even the simplest questions matter because they let the other members know I was engaged in the discussion. There were numerous occasions when something was being discussed and I had a look on my face that he knew meant there was something I wanted to ask. He'd look back at me, lower his head and raise his eyebrows as if to say, "Go ahead, Kat; ask your question!"

Over the next decade, I worked my way up to be an officer. This was huge. No other athlete had ascended to be an officer on the USTA board before. Pam Shriver had applied to be the vice president after spending many years on the board as an elite athlete, but the bureaucrats weren't ready for a former player to take such an authoritative leadership role at that time. When I was appointed vice president in 2011, I was the first former elite athlete to complete the cycle. I first had to be appointed as a director at large, after my eligibility had expired, with all of the other applicants and be respected as an individual who had leadership skills and was on track to perhaps be the president someday. I was only the second person of color to become an officer. In 1993, Dwight Mosely, a Washington, DC native, was not only the first black person to be an officer but he was the first African American to be appointed to the USTA board in its then 113-year history. He was directly appointed as the secretary in 1993–1994, and then the title was combined and he became secretary/treasurer 1995–1996. Dwight had all of the qualities and experiences needed to be the first black president, but unfortunately he lost a battle with leukemia in August 1996 at the tender age of forty-five. What I later realized, sadly enough, was that I was just the fifth African American to be a member of the board.

Being appointed the vice president was probably the most exciting moment in my tenure, even more so than becoming the president because it showed me that I was respected and appreciated for the dedication, expertise, and diversity of thought that I brought to the table. I wasn't just a professional tennis player. I was also the first former elite athlete to become an officer. This was huge and sent a message to our USTA family and the tennis world as a whole. A statement to our members, participants, and society that the USTA was walking the talk of embracing diversity and inclusion. That tennis was, in fact, a sport for all. A statement that one's ascension wasn't part of a cookie-cutter approach. That one can succeed on being different, by bringing diversity of thought from having different lived experiences than those who had the leadership role in the past. That one could look different.

I learned all the intricacies of how the USTA operated and built relationships both internally and externally with other Grand Slam board directors and executives around the world. So, when I was finally tapped to succeed Dave Haggerty as chairman in 2015, I was ready. I was later elected to the ITF board in 2015 and assumed the roles of vice president of the ITF and chairman of the ITF Fed Cup Committee.

I wanted the job, badly. But the next step was to be appointed the first vice president. It was down to me and my co–vice president, also a woman and a former section president who had a certain level of business acumen. But if the USTA was serious about being more diverse and inclusive, it was time to show it. There had never been any person of color in the top spot.

"If not me, who?" I asked the assembled nominating committee members before they cast their votes. "If not now, when?"

As soon as I said those words, I could see my colleagues shifting and sitting up straight in their seats. That top spot was mine. I had served ten years on the board to that date, giving me the knowledge and experience necessary to ascend.

I did it! I was nominated to be the first vice president—this was even more historic, more prestigious and astonishing than being the vice president. This title meant that two years later, I would be the next president, chairman, and CEO of the USTA. This was history. It turned heads and had everyone celebrating. I was attending a golf and tennis fundraiser in Florida for the MaliVai Washington Kids Foundation. It was an annual event that many former pros attended. We played a tennis round robin in the morning, and then played golf that afternoon. Well, I was in the parking lot about to drive to the first tee when the call came. It was a member from the USTA nominating committee calling to inform me that I was nominated as the upcoming USTA first vice president! Although I was hoping that I would receive this information and definitely was expecting it, hearing it was an out-of-body experience. I literally was so overjoyed that I couldn't speak. Tears came to my eyes, and I knew in that moment that history was being made—and it was about me. I immediately called my parents, and I could hear the tears of joy coming from both my mom and dad. They knew how big this was. Their black child would soon be the first black leader of an organization with a 135-year history.

This announcement was far more exciting than when I actually became the president. Everyone wanted to celebrate. Articles were written, phone calls and emails were received from people I hadn't communicated with in decades, just to congratulate me. I remember I was returning from the charity event in Florida and was flying in early that evening. My HJTEP staff had put together a celebration for me and had to have me meet them that evening, just for "drinks." Although I was tired, I agreed to meet all, going directly from the airport to the restaurant in Harlem. Well, little did I know that it was, in fact, a surprise party. All of my friends in New York were in attendance. This was a big deal for the black community. They rented out the upstairs of the restaurant and had food and drinks for all. There must have been forty or fifty people there. I was totally surprised and in complete shock that they felt that my appointment was such a big deal. At the time, I never looked at it that way. It was just another stepping stone on the path that I had set out to conquer. But I must say, this was one of the most exciting times in my life. My appointment represented an entire culture of black tennis players. People who played the sport but never saw themselves in the sport. They could now see me and believe that they now had a voice that represented their successes. As president, the number of notes that I received each time I was interviewed on television was amazing. Most people of color felt that I was representing them and that made them extremely proud to see me represent the sport of tennis. I didn't realize how much of an impact I would have on so many, young and old. Especially, each year when I had to speak, congratulating the players and giving the trophies out at the US Open. This was a moment where I felt like I was representing an entire race of people.

There was a lot I wanted to accomplish during my two-year tenure. Through the collective of great minds in my personal board behind me, urging me to stretch and grow in every position I've held, I'd developed my leadership chops to the point where I had my own style and philosophy, or what I called my "ABCs."

"A" stood for Accountability. As the national governing body for tennis, it was time for us to be more accountable for our actions in carrying out our mission.

"It's not about you and it's definitely not about me," I said in my inaugural speech as president. "It's about the brand and identity of the USTA and its role in the sport. That fact is, there is an 'US' in the USTA."

"B" stood for Behavior. Our behavior speaks volumes and can motivate others to recognize those crucial moments as opportunities to make a difference. Sometimes we need to overcome our natural instincts and make conscious decisions to change our behavior and take fresh, impactful approaches.

"Think about how your behavior toward a newcomer to the sport, whether at a clinic, mixer, tournament, or meeting will determine if that person falls in love with tennis or never plays again," I told the assembled crowd of senior staff, board members, and section heads. "The defining moment just might lie within our behavior—did we go out of our way to make the newcomer feel welcome? If we can embrace that attitude and act accordingly, our numbers can grow tenfold."

"C" stood for Communication. After ten years on the board, I realized that few outside the USTA family truly understand who we are or what we do for the sport.

"And frankly, I'm not even sure all of us *inside* the family actually know who we are," I continued. "The USTA is the most successful national governing body (NGB) of sport, not only in America, but in the world. We are, monetarily speaking, the richest organization. No other NGB enjoys the kind of proceeds generated from the US Open like we do, nor do they put their money where their mission is like we do. And our mission is clear: to promote and develop the growth of tennis . . . We have the strongest and most extensive volunteer network, including the sports' leading ambassadors. But if we're going to engage with others in our mission to grow the game and enhance lives through it, we have to do a better job of telling the USTA story."

I was determined to make an impact by strengthening our culture, and I was succeeding. But where there is that much to do, a two-year tenure can pass by in a heartbeat. Still, I was not ready for the conversation that was to come two years later. It took place just two days after all the post–US Open celebrations with champagne toasts, dinners, cocktail parties, final board meetings, and a global farewell tour as I thanked all the other Grand Slam and division leadership around the country and the world. I was exhausted and looking forward to some downtime that I could spend with friends, family, and my ailing parents back home in Chicago. A few days later, the nominating committee chairperson, Charlotte Johnson called:

"Would you serve as the president, CEO, and chairman for another term?" she asked me.

"Excuse me??" I responded.

"We would like for you to stay on as president, CEO, and chairman for another two-year term."

My head made that sound when a record player needle skids off the vinyl. "Wait, what?!?" I almost blurted. This was unprecedented. It was the first time in the 135-year history of the USTA that any chairman and president had been asked to repeat a term.

"Um, I'm here to serve the organization in whatever capacity they ask of me, but I'll need to get back to you," I said.

"Don't think too long!"

It took me just a few minutes to decide. But it wasn't just about me. There were two key people whose blessing and support were critical before I could give the USTA my answer.

My first call was to Andy Andrews, my then first vice president, who was to ascend to my role. I wanted to hear directly from him the reasons why he decided not to accept the role. Andy informed me that he couldn't continue due to family medical reasons. I was in utter shock that he declined, but I respected him all the more for recognizing that family came first.

My next call was to Roberta.

I first met Roberta Graves at the *Black Enterprise* magazine Golf and Tennis Challenge in 2000. But we really clicked two years later, at a pro-am tennis match in South Orange, New Jersey. It was a social gathering of women, some former pros and ladies who were invited by a mutual friend. When it was time for Roberta to rotate to my court, I quickly discovered that she was as intensely competitive as I was. But this was not the time or place to demolish our opponents. This was what we call "customer" tennis, where well-heeled fans get to share a court with the pros. We needed to play nice and let the amateur players win the occasional point, so that they would enjoy their experience and want to participate again in the future.

"Oh my God, tone it down!" I hissed to her between points. "It's just a game! This is supposed to be fun!" (One of our opponents was a much weaker player, virtually a beginner, and I wanted her to at least have a good time without being demoralized by the stronger player.)

"Hell, no!" Roberta yelled back at me. "If she's gonna be on this court with me, she needs to hit that ball or get out of its way. I'm the amateur on this side of the court and I want to win!" I laughed hysterically.

I got to know her better later that fall when she invited me to stay at her house in Westchester County and teach tennis to her and her kids and a host of her buddies. Given her background as the wife of the CEO of a major media empire, her impressive business credentials and vast travel experience, combined with her intellect as an honor's graduate of Yale and Wharton business school and her great intuition, Roberta was the natural selection as my executive advisor.

If I was going to do this for another two years, I needed to have Roberta's support as we traveled the world twice over, again. Given the time commitment, it was a huge ask, but she was in after getting approval from her family. She provided me the confidence I needed to own the moment.

Deep down, we both knew that this was meant to happen. It would allow us more time to effect the kind of change we wanted to see, including building up more grassroots access to tennis, developing more Latino players, leveling the playing field for women, minorities, and LGBTQ players, and seeing through the final completion of our facilities in Queens as well as our national tennis campus in Orlando. Getting all this done would have been impossible in just one term.

Fortunately, I also had a mentor within the USTA who has helped me to prepare for my career following my second tenure as president. No one knows everything. Even a CEO needs advice for transitioning to that next thing, whether that's retiring or moving into nonprofit work. A mentor has some distance from the situation. Andrea Hirsch, chief administrative officer and general counsel is someone who's been there, like a past champion who's no longer in competition with you but who nevertheless understands the pressures you are facing.

Andrea has helped prepare me for the next stage of my career as a board member in the private sector. Just as she soaked up all I could tell her about the USTA from a player perspective, I've been picking her brains about good governance, including asking all the "dumb" questions I could think of about audit and compensation committees and quarterly earnings.

Andrea not only allowed me to lean on her during my tenure, supporting my ideas or not, but I also identified her as a source of knowledge to tap into within the organization. No matter what your title is, it's imperative to have someone that you trust, who will not only listen to you but will also tell you that perhaps you might want to rethink an idea or reevaluate a situation. She really gave me the confidence to know that I could get the job done and that I wasn't there alone.

Early on, Andrea helped me to develop a résumé that plays up my leadership strengths not just in the world of tennis but also across industries. She shone a light on my strategic abilities and willingness to tackle tough issues with diplomacy and determination. Having feedback from someone with her legal and corporate expertise helped me better understand where I could go next.

3

BELIEVE IN YOURSELF
WHEN NO ONE ELSE DOES

Some people say I have attitude—maybe I do—
but I think you have to. You have to believe in yourself when
no one else does—that makes you a winner right there.

—VENUS WILLIAMS

he night before I was to give my inaugural speech in Newport
Beach, California, as the newly appointed head of the USTA,
I broke out in hives.

These were not just a few itchy welts. My body and face were
covered in huge, red angry knots, as if I'd walked into a giant bee-
hive and gotten stung head to toe. It was the same reaction I get
when I inadvertently eat blue cheese, to which I am highly allergic.
But I was sure I hadn't eaten any blue cheese that day. This was a
bad case of nerves.

It didn't make sense. Having played in front of crowds my whole
life, with experience in television commentating, and numerous

public-speaking engagements under my belt, I was usually comfortable in front of a crowd. But this was different.

The following morning in January 2015, just two weeks into the job, was going to be at my first leadership meeting, where I would be speaking in front of what was known as "the Gang of 51"—a group of tennis dignitaries from around the country that included 17 section presidents, 17 section executive directors, 17 section delegates, plus the board of directors. In addition, there were committee and council chairpersons present, along with several key staff members—about 125 people in total.

For the past two decades, I'd served on the boards of the USTA, the USTA Foundation, the WTA, the ITF, and in other capacities, so it wasn't as if I were a complete newbie to the executive world of tennis. But I was about to make my speech of intention as chair, CEO, and president, cognizant of the fact that I was the first African American, the youngest, and the first former player to be in this position. It was my first speech stating my goals and initiatives to the organization for which I was now setting the agenda. In other words, it was critical to get this right.

This time I would be bringing all my experiences into a game plan that I would be presenting to an expert and experienced group of individuals accustomed to a different way of doing things. They weren't just giving me a place at the table; I was at the *head* of the table, and now I had to prove to myself, and to them, that I deserved to be there. Of course, all of this wasn't going to happen in just one speech, but a lot was at stake. How convincingly I delivered my statement of initiatives would mean the difference between bringing these leaders on board to embrace the mission, or not. Sure, they would get it done, but not with passion if they didn't share my vision.

This is not happening, this is not happening, this is not happening, I thought, as I looked in the mirror and saw those first telltale bumps. I got into the rental car and set my GPS for the nearest ER for a shot of prednisone. I sent a text to my assistant, Nellie Nevarez, to inform her of my situation. She was ready to jump into action but there was nothing she could do but try to keep me calm through the ordeal. By morning, most of the swelling had subsided but I was still stressed. I reread my twenty-five-minute speech, practicing in front of the bathroom mirror a dozen times until I had it down cold. By the time I made it downstairs to the conference room I knew I had this, but my body was still reacting to this self-imposed pressure to be perfect. As I stood behind the podium, I started sweating, then someone's cell phone rang. It was a catchy jingle that sounded a bit like Beyoncé's "Single Ladies." I stopped speaking and started dancing to it. The music quickly transported me to a comfortable place, and the whole room erupted into laughter. The pressure that I felt as the first black president was enormous. I knew that I would be scrutinized differently and held to a higher standard, just because of the color of my skin. It was crucial that my first impression was impressive and impactful in order to get the support that I needed.

The thought of doing something new or taking what you do to a whole other level can be terrifying. But I found my courage when I remembered those who came before me, like one of my biggest *sheroes*, Althea Gibson. No one laid out the welcome mat for this icon of the sport, but it didn't stop her. Instead, Althea stepped into the arena like a gladiator, facing off the most daunting aspects of institutionalized racism and widespread prejudice. One of the first to cross the color line in tennis, she was fearless in pursuing her dream of becoming one of the greatest athletes of all time. To honor those who came before me I, too, had to boldly represent.

When I was playing, I knew I was carrying her legacy forward on the court. But it wasn't until I retired from the sport, pulling back from my life on the court to survey the whole culture, that I came to fully appreciate Althea's courage and character. She had persevered despite the tumultuous path she walked, enduring domestic violence and homelessness as a child, then facing segregation when she was not allowed to play in United States Lawn Tennis Association (USLTA)–sanctioned events. When she became a world champion, she received a ticker-tape parade, but she still wasn't allowed to stay in a hotel with whites or drink from the same fountain.

There are no words for the magnitude of what she represented not just to a people, but in the United States and around the world, in fighting the battle and overcoming incredible adversity— her impact was life changing. This fact hit me with the most clarity right before my inaugural speech as chairman and president of the USTA. As Althea put it after her 1957 Wimbledon win, "Shaking hands with the Queen of England was a long way from being forced to sit in the colored section of the bus." Understanding the magnitude of the moment, I now felt how she must have felt breaking barriers and having to represent an entire culture.

There were many other moments throughout my career and USTA tenure when I felt Althea's presence. But by far one of the proudest moments of my career took place on August 26, 2019, the year after my presidency ended, when the USTA unveiled a statue of the first person of color to win the French Nationals and Wimbledon back-to-back and the US Nationals back-to-back. It was a long time coming—some sixteen years after Althea's passing. The sport Althea loved didn't love her back in the way she deserved. The history of this sport prior to the first US Open in 1968 is not talked about or revered in the way that it should be. But the one person who never stopped talking about Althea was Billie Jean King.

Billie Jean sat down with every USTA president for the past thirty years to push for some way to commemorate a player who symbolized equality, period. She first met Althea when she was thirteen, and the experience taught her that being No. 1 in the world doesn't have to look any type of way.

"It's great that you're naming the stadium after Arthur Ashe," she told the USTA in 1997. "But our real Jackie Robinson is a woman. She broke the color barrier eleven years before a man did it."

It upset Billie for years that Althea was forgotten, so she pleaded, poked, and nudged.

"Look, I don't care if it's a court or a statue; it's about being on the right side of history."

When I became president for a second time, Billie Jean was even more relentless, because she knew exactly what was in my heart. Althea had first cracked that door of opportunity open for me and for others like me; I was walking in Althea's footsteps, carrying her legacy forward, off the court. Failing to honor her accomplishments in some palpable way was not an option.

After attending a fundraiser in Wilmington, North Carolina, I received letters from young girls in the National Junior Tennis and Learning Network program, One Love Tennis, run by an old friend of Althea's, and now my friend, Lenny Simpson. He had been teaching the history of Althea and had his players watch a documentary about her. Once they understood her significance was not just to tennis but to civil rights in our country, they became passionate about seeing something done for her, "even if it's in the form of a hot dog stand," one of them wrote. That line lit a fire under me.

The problem was, we already had a stadium named after Arthur and the entire facility was named after Billie Jean—two leading icons who've done much not just for our sport but for humanity. The USTA has a policy that no other stadia would be named after a player, following the naming of the Arthur Ashe Stadium, no matter how worthy. Meanwhile, we had already promised New York City that it would never rename the Louis Armstrong Stadium, which we had inherited from the 1964 World's Fair. But a statue would be perfect. Knowing from past experience the diplomacy and patience it often requires getting things done within the organization, timing was everything.

But when I sat down for brunch in November 2017 with Billie Jean and Ilana Kloss in Harlem, a neighborhood in Manhattan, midway through my second term as USTA President, I knew what was coming.

"Katrina, it's time. I've had it. Is this going to get done or not?"

"Believe me, Billie Jean, I want this, too. But the USTA works at its own pace."

It would involve convincing the board members who were more focused on growing tennis in America and projects like the new USTA National Campus in Lake Nona, just outside Orlando, Florida, as well as the multimillion-dollar upgrade to the Billie Jean King National Tennis Center, home of the US Open. It wasn't so much that people were against the idea as it was a case of competing interests in an organization rife with protocols and requirements for consensus. A statue, however worthy the subject, always seemed to get shoved to the bottom of the priority pile in years past.

"You have to get the Althea commemoration done under your watch. This is your chance to go back to it again, otherwise it may never happen. Just tell me what you need from me to make it happen."

Billie Jean wrote the USTA Board of Directors a letter, pleading our case:

> At the age of thirteen I saw the great Althea Gibson play for the first time, and it changed my life. Throughout her career she was an inspiration to me and to people around the world. Althea was the first person of color—man or woman—to win a major championship, winning the French Open title in 1956. In 1957, she won the title at both Wimbledon and the US Championships, and repeated the feat in 1958. In all, she won eleven major singles and doubles titles and in doing so became one of the most important pioneers for our sport and one of the most significant trailblazers for equality around the world.
>
> Althea Gibson bravely broke the color barrier in tennis and beyond, and in doing so she became a major inspiration for future generations of American players, including Leslie Allen, Zina Garrison, Venus and Serena Williams, Sloane Stephens, our current USTA President Katrina Adams and so many more. Althea Gibson was "our" Jackie Robinson and we need to celebrate her accomplishments in a meaningful and long-lasting way.
>
> It is my hope that the USTA will honor Althea with a statue, a court or something permanent named after her on the grounds of the USTA Billie Jean King National Tennis Center. It is important we celebrate Althea at the same level

we celebrate the great Arthur Ashe. Together, they changed our sport and our world. Thank you for your consideration and I look forward to working with you to move this project forward.

Sincerely,
Billie Jean King

At our next meeting, Billie Jean came in for a few minutes and spoke to us. Prior to her arrival, I had communicated the importance of approving the statue, expressing my own reasoning and passion behind recognizing this icon. When Billie Jean left the room, we took a vote and it was unanimous. There would be an Althea statue at last. Although I had been leading that discussion for years, Billie Jean had made the case with such clarity and passion, helping everyone understand Althea's significance to all cultures, not just African Americans. She reminded the board that it was a woman who first broke the color barrier in tennis, before Arthur Ashe, and she needed to be acknowledged accordingly. This was about inclusion and embracing those who look different from us. I needed someone like Billie Jean to help me get this accomplished. If I had continued to push for it on my own, it would have been viewed more as a plea for a fellow African American. But my passion for this project was not about me. It was about all that this pioneer, this legend, this icon, had accomplished despite the many obstacles she'd faced.

We commissioned the sculptor Eric Goulder, who carved her stunning and imposing three-dimensional image out of five blocks of granite, based on images of her from her twenties, when she turned the tennis world on its edge. Her quote was etched beneath:

I couldn't wait for the unveiling ceremony at the start of my first US Open since retiring from my official role, when I could finally bask in the moments of the Grand Slam tournament as a fan. I was to be one of six speakers, including Eric, Zina Garrison, Billie Jean, WTA legend Leslie Allen, and Althea's British doubles partner, Angela Buxton, who had also faced prejudice from the tennis world in the form of anti-Semitism. At one point she said out loud what many of us were thinking:

It's about bloody time!

As I crafted my speech, I thought about the time I first met Althea. The first and only time I met her I was nine or ten years old, while participating in a clinic that she hosted in Chicago. I admired her accomplishments—even if I didn't fully understand the significance of them at the time. I wish I could remember the exact words she said to me. I wish they were etched in my brain. No doubt it was something supportive and encouraging from a pioneering icon to the next generation. When you're that young, you don't always realize the enormity of the moment while you're in it. What I will never forget, though, was the way being in her commanding presence made me feel. Inspired. Whatever I was going to do in my life and career as a player, it had to somehow involve the continuation of Althea's work breaking down barriers for women and people of color. She paved a way for too many to count but not enough to say.

My mom was there the day I first met Althea. Even though my mother wasn't familiar with the sport during the peak of Althea's career, she had lived through the Jim Crow South. Because Althea had been a beacon of light for all African Americans back then, she knew exactly who she was.

Civil rights were a part of the ongoing conversation with my parents, although it was more through showing than telling. They never allowed fear to stop us or get in the way of developing an awareness of different ways of life and experience. Mom and Dad were both from Mississippi, and they made a point of taking us down there every summer to learn our roots and get to know our extended family. We rode the horses; played with the farm dogs and cats; fed the cows, pigs, and chickens; and had a blast getting to know our cousins in the South and experiencing a way of life beyond urban and suburban Chicago. My brothers Myron and Maurice were left behind many summers to assist our paternal grandfather with crops that he grew and then sold to multiple communities and stores. We zigzagged north, south, and coast to coast in our dad's Cadillac as part of our education. Nothing was glossed over. They broke it down for me and explained what might or might not happen in certain tournament locations, but never in a way that could kill my curiosity about a new city or state.

As I got older and the Chicago education system started teaching black history, I began questioning things more. In middle school I was booked to play a tournament in a state that still flew the Confederate flag. I told my mother I was pulling out.

"I don't want to go to a place where everyone can't be happy," I explained. "What kind of tennis do you think I'm going to play if I have to look at a symbol of slavery?"

"Never let that red flag stop you," she said. "I'm not having you miss out on things just because of that awful word ignorant people say. Don't give up right here because of that nonsense."

Those early lessons made me realize that it's possible to be the change you want to see, and I've made it one of my life's missions to promote diversity, inclusion, and gender equality in tennis ever since. I'm in this position thanks to my parents, as well as to Althea, upon whose shoulders I stand. That "red flag" certainly didn't stop her.

In many ways, my mother was more excited about shaking Althea's hand than I was. It so happened that my mother had passed unexpectedly just four days before I had to give my speech. So, I was speaking for both Mom and myself when I took the stage:

"Today, we honor Althea's journey, we honor her success, we honor the path that she paved for me and for all other persons of color—including the great Arthur Ashe—by unveiling this incredible monument in the shadow of the stadium that bears his name. . . . A monument that will honor the courage and commitment of the great Althea Gibson for generations to come."

It takes a level of fearlessness to make that kind of history. Winning takes boldness, something I've known instinctively since childhood.

I've been described as a player who is "aggressive at the net." My first years of learning the sport were spent on concrete courts with wire-fenced nets in Garfield Park. Often, if you hit the ball in the net too hard, it would get stuck in the holes. That always brought a laughing reaction from the other players. The indoor season started at the Old Town Boys Club and the Washington Park Field House, where I played on their gymnasiums' basketball courts, both of which had painted lines for a tennis court. These improvised courts were slick, making it difficult to rally, but that's what helped me develop my serve-and-volley game.

I wasn't interested in letting the ball bounce too many times because the "courts" were superfast. When the surface is slick, there's nothing for the ball to sink into, grip onto, and it would skid when it bounced. This was ideal for a big hitter, whose ball would hit the surface and take off in flight. As a counterattack, my preference was to serve, then move forward to the net, take the ball out of the air, hitting a volley. The aim was to put pressure on my opponent right away and end the points quickly. If they didn't have a strong return, they were usually toast. Back then, in the 1980s, serve and volley was a more common style of play among male players, although Billie Jean King and Martina Navratilova were well known for their confidence at the net. I also enjoyed watching John McEnroe, Stefan Edberg, Hana Mandlíková, and Pam Shriver, all strong serve-and-volley champions.

Most players prefer to stay behind the baseline, knocking the crap out of the ball hitting ground strokes. But going to the net requires you to see what's happening in all areas of the court. You can't hold back. You need to be fully immersed in the moment and hyperaware of all that's going on around you, staying in constant forward motion, with the ability to stop on a dime and pivot in the direction that the ball is moving. It takes hunger to be a serve-and-volley player, along with a sense of aggression and athleticism.

That's how I try to approach everything in life. First times are exciting. It's when you get to learn and experience something brand new. Standing on the brink of something is the kind of moment I live for, even when I break out into hives. That quickened pulse, those butterflies of anticipation, remind me what it feels like to be alive—similar to the feelings I would get at the beginning of a match. There's only one first time, then it's over, and I might as well enjoy the moment while giving it all I've got. I may not be successful, but if I don't put myself out there and try, I'll never know what could have been. I'm more afraid of missing out on something good than I am of putting myself in a vulnerable situation.

We live in a world of possibilities that's open to everyone, but to get the most out of every opportunity that comes our way, we've got to be bold. Particularly as women, we have to own the situation and rise above the obstacles.

I was blessed to have a family that instilled confidence in me from the moment I showed any kind of will of my own, which, I'm told, was pretty young, as I was walking at eight months. They gave me a safe place to try and fail free of judgment. If I got it into my head that I wanted to try something, my parents encouraged and enabled me, as long as it wasn't off the right path. I owned my opportunities.

Not once did I see our parents argue in front of us kids. Dad yelled at his children, but never at my mother. They were always a unified front, discussing everything calmly and thoroughly over the kitchen table. We didn't have the kind of money needed to play tennis growing up, but as a child I never felt deprived because of all the careful budgeting and sacrifices Mom and Dad made for us. Without their teamwork, my career in tennis would never have been possible. Mom held down the fort with my two brothers while Dad was on the road, taking me to tournaments.

They had an unspoken understanding, each one always knowing exactly what was needed from the other. If we needed to buy me new tennis outfits, Mom would cut back on certain things in her monthly budget, although she also saved money by sewing some of my clothes. Dad managed the home budget and banking, but as conscious as Dad was about every penny that was being spent, he never micromanaged my mother. He trusted her to make the right decision for our family, just as she trusted him.

Being around my brothers and my cousins, who were mostly boys, I was always around the big kids playing sports, from softball in our backyard to throwing a football to playing strike 'em out.

I wasn't about to be left out of the fun they were having simply because I was a girl. By the time I came along, my oldest brother, Myron, whom we called by his middle name, Andra, was almost old enough to babysit. Since my parents both worked, the rule was that my brothers had to allow me to tag along whenever they had neighborhood activities outside of school. The catch was that I had to be decent enough with a ball not to embarrass Andra with his friends (Maurice, my other brother, was another story). Being the ultimate tomboy, I was.

Mom recalls first seeing me playing with a tennis ball in the backyard when I was no more than two years old. At that age I was drawn to it because the ball was soft, fuzzy, and small. We had all types of balls around the house, small and large. But tennis didn't really come onto my radar until I was six. Even at that young age, I identified with Arthur Ashe, a man who looked like he could have been one of our family friends.

"One day I'm going to England to play," I announced to my mother as I watched a member of British royalty present him with a gleaming trophy. She just smiled.

When my brothers got to join a police-sponsored tennis program at the Dr. Martin Luther King Jr. Boys Club in Garfield Park a few weeks later in the summer, I had to find a way to get in on the action.

The park was about four blocks from our home. I sat on the sidelines for the first couple of weeks and watched the older kids play. Maurice and Andra, who were thirteen and fifteen, respectively, made it clear they did not want to be there, and their bad attitudes were getting on the instructors' last nerves. One of the coaches was making Maurice do ten laps around the park and he was not having it, while Andra was being his usual cocky self. The other kids were hitting balls all over the place, such as into the net or over the fence. I remember wriggling on the bench, feeling impatient and thinking, *I can do that.*

Kim Williams, a young assistant coach, must have caught the look of eagerness on my face. Either that or he'd taken pity on the little girl sitting there all by herself, staring pathetically through the chain-link fence. I had been asking the program supervisors for two weeks to let me play, but I was too young. Kim finally called me over and invited me to give it a try, showing me the basics. Then he put the adult-size racket in my hands, showing me how to grip its handle, where to place my feet, and how to rotate my body as I hit and returned the ball.

Kim was dumbfounded when I was able to perfectly replicate every move he showed me. So were my brothers, who couldn't hit the ball to save themselves. I was a visual learner, and after watching the other kids daily, I knew exactly what I was supposed to do. Soon it seemed like the whole park was watching as I rallied back and forth with Kim. No one could believe how quickly this pint-size kid was picking up the sport.

At the time I had no idea that this would be my destiny. But there was something about the feel of the ball on my strings and hitting it over the net, into the court. From that moment I fell in love with that sense of mastery and control of the court that I wanted to experience again and again. I was always intensely competitive, so I loved being able to show them all. From that point on, I just wanted to get better and better. Young as I was, I was fired up about learning everything I could about tennis.

When I got home that evening, I told my mother yet again that I wanted to join the program. It was only supposed to be for kids aged nine and up, but Mom didn't realize the coaches were willing to make an exception for a six-year-old.

"But honey, that's for the big kids," she told me.

Arms folded and indignant, I dug in my heels.

"I'll be seven soon, and they're not good enough to play with me!"

It helped to have the program's head coach, Tony Fox, in my corner. He believed so much in me that, when he saw my older brother Andra was acting up and about to get himself thrown out of the program, he asked the lead instructor to keep him in just so that I could continue to play. Tony knew there was no way I would be able to come by myself. He even went over to my house to have a talk with my mother, promising her that he'd personally see to it I'd be brought home safely at the end of each practice.

"This child is a phenomenon," he told her. "We've got to find a way to let her keep playing, because she's going to go far."

That conversation changed everything. He spoke with such passion, sincerity, and authority. Thanks to Tony, my mother understood tennis was going to be in my life from that moment on. My two older brothers were under marching orders for the rest of the summer. They didn't have to play or participate themselves, but one or both would have to get up each morning, walk me to the park, sit down, watch me practice, then bring me home when I was finished. For the rest of the summer Andra and Maurice were the ones sitting on the sidelines while I got taken under the wings of Kim and Tony. And if for whatever reason they weren't there, Tony, or one of the other instructors, would escort me home.

I became so obsessed with the sport that Mom had to cancel my ticket for the mother-daughter educational trip to Europe she'd planned with a group.

"Tennis is here, not over there," I told her. "Please let me stay!"

My parents had some conditions.

"Books first," Mom told me.

If I was serious about tennis, I also had to give it my all. I kept my part of that deal. Even at six going on seven I knew then that I would need to be disciplined, taking every opportunity I was given to train. My family was middle class, both parents were educators and homeowners. They didn't know anything about tennis, but they could see how passionate I was, and they supported me and made sacrifices. We couldn't afford the airfare to get me to tournaments around the country, so my dad and I piled into the car and drove for hours. But it was up to me to find every opportunity I could to get better.

I don't put much stock in beginner's luck. Being good on my first try at tennis was no fluke. My whole life I'd been watching my oldest brother, my male cousins, and all their friends play sports. Andra used to wrestle with me to make me stronger and tougher. Playing softball also gave me great hand-eye coordination. By six I could throw a football with a spiral, which gave me the muscle memory for the service motion. There were multiple skills I had already developed at a young age that translated straight into playing tennis. By the end of that summer I was given a little trophy for "most improved player," which I've taken with me everywhere I've lived.

A sense of fearlessness drove me. When I was fourteen, I challenged this tall dude with dreadlocks, Ivan Daniels, who used to dominate the courts. Known as one of the local tennis bums, Ivan was by all accounts a strong amateur player. One day I asked him to play with me. He lost in three straight sets.

"I just got beat by a girl!" he wailed as he walked off the court.

According to Ron Mitchell, a family friend and prominent member of my local Chicago Prairie Tennis Club community, in 1984 I played the first round of a match in one of the affluent suburbs of northern Chicago entirely with my left hand.

"You were just showing off and being one of the baddest dogs on the block," he told me. "I was scared witless you were going to clown around and lose this."

I don't quite recall it that way. Being the only player of color at that tournament may have been part of the reason for my strut.

Off the court, I rarely misbehaved as a kid. On the court was where I worked out all my aggression. Being the first to move forward to the net put me in control and exerted pressure on my opponents. But I had to do the work before I got there, choosing my moment, mastering my footwork, which always needed improvement, and setting things up by hitting hard and sometimes high, deep balls to get to the net. Never let self-doubt weaken your stroke.

My decision to turn pro had not been made lightly. With parents who had always stressed the importance of getting a college degree, leaving Northwestern before graduating wasn't necessarily part of my original game plan. My parents and their siblings had earned their college degrees in the South. It is very important in our culture to not only attend college but get your degree. I was choosing to forego this in a typical four-year timeframe, perhaps finishing later instead, if my career wasn't successful. That hasn't happened yet. My whole life playing local tournaments on weekends was simply about walking away with the trophy. In the 1980s, most tennis players with potential turned pro after their freshman year, if they went to college at all. But I was enjoying my classes, the campus life, and the fact that I was crushing it at the collegiate level, having made it all the way to the semis of the NCAA Women's Doubles championship my freshman year, in 1986. My doubles partner Diane Donnelly (Stone) and I were setting records, with twenty-six victories in 1986 and twenty-one consecutive wins in 1987, culminating with the NCAA Doubles Championship.

But by then I'd landed on Billie Jean King's radar. When she was living in Chicago this icon had taken it upon herself to give me a crash course in fearlessness.

My college coach, Sandy Stap Clifton, called BJK to ask if she would practice with me at the Midtown Tennis Club, where she and her partner, Ilana Kloss, played whenever they were in Chicago. Sandy knew she would say yes, because Billie has spent her entire career championing and mentoring young female players. Billie later told me how she liked the way I kept coming forward on the court. I got a kick out of meeting and playing with the legend, but I filed away the excitement of that moment to focus on practice, and homework. A few weeks later the phone rang in my dorm room. I was down the hall playing billiards with the boys. My roommate came running to get me with a panicked look on her face.

"It's Billie Jean King!" she screamed. "Billie Jean King's on the phone, and she's asking for *you*, Katrina! Come quick!"

Billie Jean had convinced Zina Garrison to invite me to practice with her and Lori McNeil in Houston that winter, during their off season, as they prepared for the Australian circuit. The gift of that experience was a turning point in my playing. It gave me a true sense as to what it meant to train like a pro. Zina and Lori were among the top ten players in the world. They owned Grand Slam Doubles and Mixed Doubles titles and had been to the quarterfinals in singles of Grand Slam tournaments. I had never worked as hard as I did under the hot Houston sun.

The workouts were triple the intensity of my previous training, on and off the court. After two or three hours of drills on the court, I had to run the bayou with Lori, who was built like a racehorse and ran like one too. She never slowed down or ran out of gas, whereas I must have thrown up every day for my first three days there. I was coming to grips with the fact that I was not in shape, at all. I guess the late-night beers and pizzas in college were taking more of a toll than I realized.

This experience gave me an inkling as to what it would take to be a professional tennis player, and it took me to that next level as a college player. I started my next term as a sophomore fighting fit. After winning the NCAA doubles and being voted all-American twice, I would have started my junior year ranked No. 3 in singles in college. According to Sandy, these wins were signs I was ready to go professional. But to stop short of obtaining my degree and risk upsetting my parents, I had to be completely sure I could make the jump with no regrets. This was a big deal for my family, because a college degree had always been regarded as my passport to financial security. It's not as if we had some sort of financial safety net if I walked away from my scholarship, then failed to earn a decent living at professional tennis.

I promised Mom and Dad I'd finish my education; how can I do this to them? I asked myself.

It was always my intention to complete my degree; even now. Neither my parents nor I realized that, at that time, it was standard practice for most junior players to turn pro in their first year of college, so I was already late. But the prospect of letting my family down still sickened me.

I did what I often did when I was facing a crossroads or about to embark on some major challenge. I made a mental list of pros and cons, then talked to my coaches and mentors, figuring out all the possibilities and potential pitfalls on the path I was about to take.

Eventually, I decided to take the fall quarter off from school to play as an amateur. For the past ten years of my life I'd traveled and played tournaments all day, every day when school was out. I wanted to see whether that would feel like a grind outside of summer or if I enjoyed it.

"Let's see how it goes," I told Sandy. "If I don't like it, I'll come back to school, keep my scholarship, and we'll never have this conversation again. If I do well, then maybe I'll turn pro in January, but at least give me the opportunity to test things out." I had an amazing experience at Northwestern. Although I was the only black player on the team, that didn't bother me, especially since I was one of the top players. In situations like that racial slights are uncommon. My teammates were great friends, and I didn't feel any tension. But it did help knowing that I had my nucleus of black friends that I returned to after practice or road trips, where I could be fully myself and didn't have to be conscious of "representing the race." I always looked forward to being in that comfort zone.

I was realistic about what success would look like. Turning pro with a ranking of 200 or 300 wasn't going to get me far. But at the end of that year, 1987, I was ranked 104th in the world and qualified for direct entry into the Australian Open. This time I felt ready to launch myself on the professional tour, improve my ranking, and support myself financially. My parents understood. When we finally had the conversation, they could see the conviction in my eyes. Looking back, I'm proud of the path I'd chosen: to allow myself enough time to mature as an individual at Northwestern and reach a level where anything was possible. I was as prepared as I could be as a rookie, with every chance of making it to the top ranks of the sport.

With the level of training that's available to young players today, reaching the top right out of the gate is even more of a possibility. Consider the 2019 US Open Women's Singles champion, Bianca Andreescu, a nineteen-year-old Canadian of Romanian descent who was ranked 152nd before she beat Serena. Bianca, who had battled her way back from injury, became the first Canadian ever to win a Grand Slam tournament singles title, despite her not even qualifying for the Grand Slam the previous year. Coco Gauff is another shooting star who, at fifteen, upset Venus Williams in the first round at Wimbledon in 2019 and became the youngest player ranked in the top 100 by the WTA with a ranking of 68 in the world.

My own launch onto the international tennis stage was less spectacular. There were certainly plenty of distractions at the 1988 Australian Open in Melbourne. It was in January, the peak of summer Down Under, where the heat got so intense at times that we players could feel the burn as we drew air into our lungs. Despite the temperatures, I fell instantly in love with the city. The fact that the facility was walking distance from the hotel downtown where I was staying allowed me to soak in some of the local life. When I wasn't practicing or competing, I filled in the hours strolling through the streets and listening in on the conversations, completely charmed by that distinctive Aussie accent. Everything was familiar, easy yet different somehow, and slightly discombobulating: from the fact that gasoline was called petrol and sold in liters to the Aussies' taste for Vegemite, which I never acquired. But my attraction to the place was more about the personalities of the Australian people, who were fun, easygoing, and relaxed. While not exactly ethnically diverse, I was hooked because they treated me like I was one of their own. However, they had multiple cultures—Italian, Greek, Chinese, and other Asian cultures. This offered great restaurant choices for different cuisines. I quickly befriended three families, thanks to Zina and Lori, who had befriended them years ago! A colored South African family, an Italian family, and a Greek family, all who understood the importance of family and welcoming you with amazing food. They truly made me feel at home, and I embraced their diverse backgrounds and values. That's so important to receive when you're a minority far away from home.

I also relished the charged atmosphere of the stadium in Flinders Park, now Melbourne Park, where I was surrounded by some of the greatest players in the world: Chris Evert, Steffi Graf, Pam Shriver, Martina Navratilova, Mats Wilander, Pat Cash. I'd worked toward this moment my whole life, and now here I was. There wasn't as much pressure on me because I was not expected to win, but I had flown all the way to the other side of the world to play in my first Grand Slam tournament. I did not come this far to lose. In my head, this tournament was about justifying one of the hardest decisions I'd ever made. Maybe that's why I let my emotions get the better of me during the first-round singles match. That Melbourne was debuting its brand-new stadium should have been a good thing, but its state-of-the-art hard, high-bounce court surface didn't exactly give me the edge I might have enjoyed if they'd stuck to the old grass courts.

It all added up to the loss of my first professional singles match. My opponent, a fellow American, Marianne Werdel, whom I knew from our years playing as juniors, beat me in straight sets. She was a couple years older than I was, more experienced, with a huge serve and forehand, but I believed I would win. Doubles didn't go much better. Although my partner, Penny Barg (Mager) and I won the first couple of doubles rounds, we lost in the third to two icons: Chris Evert and Wendy Turnbull. Australia was a bust! At least that was my interpretation, not having left with a trophy.

My debut professional performance amounted to a humiliating pummeling, fortunately on a side court outside the main stadium and not in front of a large audience. For a moment, it made me ask myself, *Do I really belong here?*

It's easy to get intimidated when you are brand new to a situation and surrounded by titans. You want desperately to prove to them that you deserve to be in that space. But having that experience also allowed me to understand all the external pressures that professional players experience in these large arenas, where the whole world is watching. Instead of dwelling on the loss, I decided to savor the moment I was in. I got to occupy the same space as the greatest players in the world whom I had admired and looked up to my entire life. There I was, in the same training room getting pampered by the WTA Tour medical team, getting my ankles taped by the physiotherapists and receiving post-match treatments alongside Gabriela Sabatini, Helena Suková, and other players. I could get used to the pristine locker rooms stocked with toiletries worthy of a five-star hotel. Carb loading on pasta in the player restaurant never got old either, especially when I got to sit across the room from idols like Yannick Noah and Stefan Edberg, and their entourages.

Some of the magic dust must have rubbed off, because it was just enough to keep me moving forward on the tour. Under the tutelage of my coach, Willis Thomas, I continued to develop my skills, stick to my game plan as a serve-and-volley player, and get myself into better shape overall. Willis is black and was a huge asset in my career. I believe having a black coach gave me the confidence that I needed, since he came from the same culture and understood the challenges and biases that were out there. His presence gave me a comfort that I needed at a young age to allow me to be me, not feel like I needed to be someone else to fit in. Plus, I have observed that black players who have had a coach of color at some point in their career perform better overall.

When Willis retired, I then worked with Charlton Eagle, a colored South African, who was Australian. I needed that same type of chemistry and he traveled with us a few years prior, as a hitting partner. There's a different, higher level of belief instilled in the player. Because of the long history of racism in America, I think it is challenging for whites to see blacks fully and to see our full potential. And, in tennis, prior to Venus and Serena, it was assumed that because tennis is a mental sport black people would never have a strong presence in the game. But the good part about stereotypes like these is that reality will always shatter them eventually. When I returned stateside, I started winning more matches. Before the end of that year, I'd reached the fourth round of Wimbledon, losing to Chris Evert, and the semis of the doubles. I'd also won the World Doubles Championships in Tokyo. Along the way, my singles ranking climbed from No. 104 to No. 67 in the world, and my doubles ranking went from No. 105 to No. 14.

I kept climbing, finding my rhythm on the professional circuit, especially as a doubles player. Through it all, the hectic pace and travel of that life felt like an adventure. From one tournament to the next I never knew what lay ahead, but I did my best to enjoy the journey, soaking in the culture whether I was playing in Japan, England, or France. Although I didn't have much time to sightsee, I felt it was important to appreciate the difference in cultures and respect their customs. No matter how prepared I was for the next match, the next career change, the next big moment on the podium, I took nothing about the world I was in for granted, knowing that in my sport, things can change on a dime, whether from injuries, losses, or a winning streak that got me into the final rounds. Whatever was going on, I left it all out there on the court, playing with a kind of go-big-or-go-home abandon.

However, life wasn't always easy. I lost more than I won, and the stress of making money to pay bills was very unnerving if I was having a losing streak, week after week. Year one of the tour can be your easiest because you don't have major expectations. But after a somewhat successful rookie year, it got tougher because I had expectations, and maybe, just maybe, I took things for granted. When winning wasn't as frequent as expected in year two, I struggled emotionally and had to refocus. That's when I began to train harder on and off the court and focus on the present instead of on the dream of being a champion. I had to do the work to be a champion over and over again. No one was just going to give me a match because I might have beaten them the last time we played. They were that much hungrier to win the next time. I had to learn to prepare better for every match and learn from the losses. Sometimes you can get complacent when winning is frequent. I had to stop relying on just my talent and rely on the complete package: talent, fitness, tactic, and overall preparation!

The mental and physical toughness was necessary, because I came of age as a professional in the era of power players like Monica Seles, Jennifer Capriati, Lindsay Davenport, Mary Pierce, and the Williams sisters. Their strength pushed me to play better, until the injuries slowed me down. I retired the year Serena won her first Grand Slam tournament singles title. In fact, in February 1998 at the IGA Tennis Classic in Oklahoma City, she was my last singles loss. Serena was seventeen and I was twenty-nine.

In this moment, I recognized that the next generation of players was solid and they would be the future champions of our game. I was proud to see this young, strong, and powerful black teenager solidify herself in our sport. She had yet to win her first Grand Slam singles title though that would come later that summer. But for me, I was thrilled to see the next generation of black players making their move. I had played Serena and her sister Venus in doubles and felt the force of their power, but to be on the court playing singles was a whole other experience. One of feeling exhilarated and exposed at the same time. Exhilarated knowing that Serena was the future and exposed in knowing that I no longer had the game to compete with the next generation. These are moments where it's important to recognize this changing of the guard and lift up and support your successors no matter your industry. Take ownership in knowing that you were a part of their journey to success. I knew at that moment the sport I loved so much was passing me by and that I was no longer the same force. I became a doubles only player for the next two years.

Halfway through 1999, I began coaching for a while, as that is what many pros do after they retire from professional play. It was rewarding on many levels, but eventually I decided to enter another arena and make the career switch from coach to television commentator. Given the confidence I gained from tennis, I wasn't about to let lack of experience get in the way of taking my best shot. As women, we tend to hold ourselves back more than we should. Studies show we don't apply for jobs unless we meet 100 percent of the requirements, whereas men have the swagger to go for it even if they only tick 60 percent of the boxes. We women tend to bog ourselves down with self-doubt. We stick with what we believe we're best at and often avoid the risk of change. But I had too many pioneering women in my life, like my mom, Billie Jean, who has always encouraged me to go for it, and Lesley Visser, a true badass who broke every barrier she could think of as a sports journalist.

I first became aware of who Lesley was in my debut year of turning pro, when the media was buzzing over my early success at Wimbledon and the short-lived scare I gave Chris Evert when

I played against her in the fourth round. Lesley, who was covering the event for CBS at the time, jokes that I blew her off for an interview, although I don't exactly remember it that way. But now, as we've gotten close as friends and colleagues on the first all-women's sports talk show, *We Need to Talk*, I look back on her career in awe.

The most acclaimed female sportscaster in history and the first woman to cover the NFL as a beat, Lesley was also the first woman to be recognized by the Pro Football Hall of Fame, the first female anchor on ABC's *Monday Night Football*, the first female sportscaster to carry the Olympic torch, and the only sportscaster—male or female—to have done the network broadcasts of the Super Bowl, the Final Four, and the NBA Finals. If it's a sport, Lesley has covered it, whether it's the Triple Crown, the US Open, or the World Figure Skating Championships. Voted the No. 1 female sportscaster of all time by the American Sportscasters Association, Lesley is everywhere! But access didn't come easily when she was first starting out in the business. The job of sports journalist didn't even exist for women when she was growing up in the '60s. With women like her blazing a trail, who was I not to follow?

Being bold got me a foot in the door at the Tennis Channel in April 2003. Although I had only ever seen two black commentators covering tennis on television—Arthur Ashe and Zina Garrison—I was still determined to make it happen for myself. And I knew it was on me to prove them right in their decision to hire me. That first moment on air was the bigger challenge. Sure, I'd studied communications for two years at Northwestern. I'd been around reporters and cameras my whole career as the subject of interviews, talking about a match I'd just played. I even had the chance to be a "guest" analyst in the booth at some events for a few matches. But this was a new kind of pressure. When that on-air light flashed red, I had to be on point, natural, and watchable. Just as the best tennis players make it look easy, live broadcasting is nowhere near as easy as it looks.

This wasn't just my first time performing as a color commentator. It was the Tennis Channel's first time going live. Oh, and did I mention it was live?! That meant I could not stop tape whenever I flubbed. There would be no retakes. I had only one chance to get it right. I also knew that I had to get it right and not screw up because I was Black. I felt that moment was not only critical to me having more opportunities to do more events but to give hope to other black players that they too could be a commentator. Remember, I had only seen two before me, a decade apart. I also knew that I was making other black viewers proud to see someone who looked like them. Extremely proud. I was representing all. Unless you're in my skin, in my shoes, you don't know what that feels like. Unless you yourself are the only one in your arena.

We were doing a broadcast of the Fed Cup in Lowell, Massachusetts, where the US was playing against the Czech Republic. The rehearsal went well. But then it was time for the countdown. As I stood next to veteran play-by-play analyst Barry Tompkins my knees were shaking, and my teeth were chattering so loudly I was convinced people could hear it clear across the stadium. I was self-conscious about it and made a point of loosening my jaw so my teeth wouldn't touch. Barry started talking, introducing the show to our viewers and saying a little about that day's tennis. His words felt distant, remote, as if I was watching him as an audience member in a movie theater. I remember thinking, *Wow, I can't believe I'm here . . . wait, I am* here. *Oh my God, this is really happening!* Then it was over to me.

Barry asked me a question about the match. I knew the camera shifted to me, remembered to smile, then started talking. It was such an out-of-body experience that I have no recollection of what I said, although, apparently, I *was* coherent, because the producers seemed pleased. I relied on my notes, my knowledge of tennis, and the earlier prep work we'd done. There was no time to freeze; letting people down wasn't an option. I only had a few moments to say something impactful and relevant. About halfway into minute two of the three-minute opening segment I began to ease into it. The camera was now on the match, not me, so it was just a matter of talking about what I was seeing. Before I knew it, that first time on live TV was over. They say never let them see you sweat. I didn't, but believe me, I was drenched under my suit jacket.

I was pumped by being challenged with the unfamiliar. The pounding heart, sweating palms, and butterflies remind me of the joy of new experiences. Knowing I'd gotten through something and performed well made me want to raise my fist in the air and celebrate. Even if I don't nail it on the first try, it's an accomplishment. I add it to a long list of firsts that have guided me onto some new and exciting path. When I face my fears and do something completely new, I'm showing up for my life. I believe I'm stepping into the arena and becoming the hero in my own life story. As Brené Brown said, "If you're not in the arena also getting your ass kicked, I'm not interested in your feedback." Whatever the challenges, I owned the experience.

This is how I approach my life. Although I may be the only black person in a room, I silence any self-doubt. I own the room with confidence, convincing myself that I belong there and acting like it from the moment I enter. At times, it can be tough to pull off, but my parents taught me that I belong wherever I am. Naturally, the discipline and style of play I learned in tennis has carried over into my professional life. At the exclusive level of sports management,

I'm often the only black person in a variety of situations, which was similar to my tennis development where I was the only one in the draw. To have presence in these rooms, frequently filled with white men, I rely on my life skills gained through tennis to make a meaningful contribution. I learned this early on, when I joined a new board where I was the only woman sitting at a table of nine men. I shared some insight, I can't recall what exactly, but the response has stayed with me ever since:

"Thank you for sharing that wisdom, I hadn't thought of it that way," one of my fellow board members said.

"Of course, you wouldn't," I told him. "You're not a woman!"

It was a real V8 moment for all of us on the value of diversity of thought.

But you still have to bring it. First and foremost, I always make sure that I am prepared to discuss whatever issue is at hand. Commanding the room is not an entitlement; it's the product of being prepared. I bring my knowledge of a subject, as well as my creativity. God has made this particular place for me, and it's up to me to rise to the occasion by demonstrating my talent and worth, knowing that I often represent more than just myself.

4

EVENING THE SCORE

The day I stop fighting for equality . . .
will be the day I'm in my grave.

—SERENA WILLIAMS

We were all waiting in eager anticipation for the 2015 US Open Final between then No. 1 ranked Novak Djokovic and the No. 2, Roger Federer, but the weather wasn't cooperating. This was going to be an intense matchup between two of my favorite players. Adding to my excitement was the fact that this would also be my first Grand Slam men's tournament final as head of the USTA. But each time we thought the first game could start, the heavens would open and turn Arthur Ashe Stadium into a small lake. Of course, this had to be the year before we completed the installation of the retractable roof.

The three-and-a-half-hour delay was a drag for the fans waiting around in the stands. But it was a logistical nightmare for the President's Suite, which is a revolving door for fourteen days, hosting anywhere from 400 to 450 people a day between two sessions, and on finals weekends up to 220 people for one session. The process begins in early May, when we request the invitation lists from various board members, senior staff members, sponsors, and so forth. Those requests are broken down by which day and session each guest will be attending, so that we can then figure out all the other logistics, such as who sits where, and next to whom, whether they are my personal guests, international guests, board of directors and their guests, sponsors, or other special invitees. None of the groupings are random. We want the conversation to flow as people make exciting new connections and leave with a renewed sense of passion and commitment to our sport. We want the whole experience to be memorable. To that end, we put a lot of thought into the wine lists, menus, table linens, place settings, and flower arrangements. When we sit down to discuss our vision and culinary preferences, no detail is too small.

We followed the same scheduling for our guests as in previous years, including that morning's Sunday International Brunch, where we give speeches and toasts of thanks to dignitaries from around the world. After this lavish meal, it's customary for us to serve sandwiches and finger foods like crab cakes and curry puffs throughout the rest of the day, along with beverages and cocktails from our fully stocked bar. It's plentiful and almost never runs out for the next round of guests, who start to arrive around 2:30 p.m. for the 4:00 p.m. match. But with the intermittent rain causing our VIPs to stand around with nothing to do but drink and linger in the lounge area until 7:00 p.m., we had a major hospitality problem. The room was getting crowded and warm, with two hundred people, who couldn't go outside without getting wet and ruining their immaculate outfits and hair. The snacks were running low, and our celebrity guests, who'd lingered since the brunch to enjoy the match, were starving.

More than one of our VIPs needed some carbs to soak up the alcohol from a steady stream of mimosas, champagne, and honey deuces. A few of them, including longtime US Open friends Mariska Hargitay and Hugh Jackman, quietly came up to us and politely asked if they could get something to eat. Because they were high profile, there was no way they could easily slip out to the food concessions without attracting a crowd and creating security challenges. Besides, after patiently waiting around for hours for the tennis to start we couldn't *not* give them a sit-down meal!

Our schedule on the men's final Sunday had always been an International Brunch, then no more chef's table for the rest of the day. Despite holding some of the most exclusive and coveted invitations of the entire tournament, these guests were getting the short end. The least we could do was make sure they got fed. Just because something has been done a certain way for decades doesn't mean it shouldn't change. So, after experiencing one too many rain delays, my team—consisting of Cathy Politi, a USTA veteran of twenty years and head of the President's Suite, Roberta, and my executive assistant Nellie Nevarez—later decided that the timing of our meal service had to be changed. We'd make the brunch earlier, then lay out a substantial late lunch / early dinner to fuel our guests no matter how much it rained or how long a match would be played.

Of course, that didn't help our starving VIPs on this occasion. We needed to come up with a plan. Fortunately, that year we'd rehired an executive chef, Jennifer Cox of Levy Restaurants, to serve as chef of the President's Suite, and she is as resourceful as she is creative in producing the kind of food that pleases the most discerning palates. When we asked her what she could do for us, she looked in the fridge and found seven leftover servings of the brunch that she could whip up and repurpose into an early dinner for our ravenous celebrities. But where to serve it?

The dining area was closed, the tablecloths and settings long since packed away and locked up. Most of the tables had been broken down, although a few remained standing. In any case, opening the room would have signaled to the rest of the guests that we were ready for business when there simply wasn't enough to go around. As a last resort, Chef Jen proposed a discreet "chef's table" in the kitchen. She was pleasantly shocked when these actors loved the idea. She then kicked into high gear, clearing off a table and setting it with some glasses, napkins, and plates for an intimate setting.

The celebrities were thrilled with the arrangement and spent the next hour and a half eating a five-star meal, hanging out and taking selfies with Chef and her staff. The table wasn't even necessary. All gladly ate standing up. The spontaneity of the moment made it priceless for everyone. The flawless execution made it an invaluable lesson for me early in my USTA tenure. It was with utmost gratitude I realized that the excellence of my team made me shine like a diamond. I was absolutely thrilled when I was called to the kitchen to witness this quaint arrangement and the happy guests.

A strong, productive team requires inclusion of different ideas, abilities, and backgrounds. It takes humility and strength as a leader or champion to hear a different opinion or let someone else take the ball. To play your position well you need to collaborate and communicate. You also need diversity of people, experiences, cultures, and thought. No matter what you are trying to accomplish, the more points of view, the better the outcome, even if it takes more discussions and debate to get there. It's always worth that extra time and work to hear out and cultivate different perspectives. Members of diverse groups challenge each other to think more broadly and be more innovative in thinking and decision making, exceeding original goals and expanding beyond the original vision.

The experiences I've had as a doubles player make the strongest case for diversity. When my doubles partner was exactly like me, we rarely made it to the championship match. Lori McNeil, one of the other great players of my generation, was too much like me in temperament and playing style. We were both powerful, athletic, and aggressive players, but that got us only so far. In my opinion, that's why we didn't win as many titles. We often reached the finals but couldn't always capture them. When you have the same mentality on everything you can't see the fuller picture. You often bump heads, but having varying personalities and skills that work together helps to get you across the finish line.

I began to appreciate the importance of diversity when I started traveling globally on the WTA Tour, although you tend to stay inside a bubble of practice courts and hotel rooms as an athlete. But it wasn't until I was involved on the organizational side of the sport, first as an ITF delegate, where I met all the other international representatives of the sport at my first AGM in Copenhagen in 2012, then as the USTA President attending Fed and Davis Cup ties in Serbia, Croatia, and many other countries. In these roles I could focus more on the different environments I was traveling through, from Dubai to Singapore. The more people of widely different cultures and backgrounds I got to meet, and the more ways of life, social norms, governments, and religions I witnessed and experienced, the better I felt I understood the complexities of the world.

Beyond developing a greater tolerance for differences, it taught me a deep appreciation for the rich tapestry of humankind. Of course, as the only black person in many of these places, I stood out. But I quickly learned that the rest of the world doesn't necessarily see color as a racial barrier. It made me appreciate diversity all the more.

Of course, the topics of diversity and inclusion have been deeply personal to me from a young age. In the 1970s, coming straight off the heels of the civil rights movement, equal opportunity was still a relatively new concept. It was rare to see people of color in our sport. And I felt their absence. After Bob Ryland, the first professional African American tennis player, Arthur Ashe, and Althea Gibson, players of color were few and far between. Leslie Allen was ranked as high as No. 20 in the world, winning the Italian Open. Renee Blount, Kim Sands, Diane Morrison, Zina, Lori, Camille Benjamin, Jeri Ingram, Chanda Rubin, and a few others also gained prominence, but they were spread out over a couple of decades. Following earlier success by Chip Hooper, Rodney Harmon, and MaliVai Washington, and others, a few years later there was also James Blake, who reached No. 4 in the world despite battling serious injuries.

We all gravitated toward each other because we didn't see anybody else out there who looked like us. It was a small, exclusive club. We did have a ray of light when Yannick Noah won the French Open in 1983 to become the first black man to win a Grand Slam title since his idol Arthur Ashe. But he's not American; he's French.

It still bothers me that, more than three decades after I first picked up a racket, kids of color are still a rarity in the sport. Tennis is still struggling to diversify, with 77 percent of participants in youth tennis made up of whites, compared with 9 percent African Americans and 14 percent Latinos. The reasons for this situation are many, although it's mostly due to a lack of opportunity and exposure to the sport. The only way to change this and build up a more diverse pipeline was to get more involved at the grassroots level. Having come through every part of the tennis ecosystem, with hands-on experience as a professional player, I knew I could have a real impact. As Billie Jean King likes to say, if these kids could see it, they could be it.

I was fortunate. I rarely, if ever, felt excluded. Mom and Dad did a good job of protecting me from the effects of any prejudice as a child. But it's only as I look back as an adult that I realize the fear they felt for me as I traveled throughout the country to various tournaments, especially when I headed south. I was ten when I went to Arkansas for my first out-of-state event without my parents, who both had to work. Housing was offered back then, where volunteers of the club or community offered their homes to players travelling alone. They treated us as if we were their children and made sure we had food, shelter, and transportation to the tournament. I was to be the only black kid playing in the tournament, so my mother called ahead to an old neighbor of ours who had moved there to ask what it was like. It was 1978.

"The racism here is something else," he informed her.

Panicked, Mom immediately called the tournament headquarters and explained I wouldn't be coming:

"I'm so sorry, but Katrina's not used to this; we're pulling her out."

The tournament organizers were as reassuring as they could be, promising her that I'd be well cared for and protected in the home where I'd be staying. Just to be sure, Mom called my hosts, tennis fans and members of the country club where I would be playing. They also happened to be white.

"I'm a mother too," the lady of the house told her. "I promise to protect your daughter as if she were my own."

When I returned home, my mother didn't want to directly ask about my experience, so she fished.

"How was your trip, Katrina?"

"It was fine. I won."

"Was the food good?"

"Momma, I had plenty of food, I got picked up and dropped off every day, and everyone was nice to me. Don't worry!"

"Well I'm just a country mother and it's hard for me to get used to you traveling when your dad and me can't be with you."

"I promise to tell you if there's a problem."

But there was no problem. My accommodations were no different than being at my own home. The house was nicely kept. They had plenty of space and the food was great. The only difference was that they had a Ping-Pong table, where we had a billiards table instead. I always suspected my family provided a kind of buffer against more negative experiences on the road. Whatever bigotry we experienced was subtler, along the lines of patronizing assumptions. One of my father's favorite stories to tell occurred at a country club in Alabama. The father of one of the other junior players on the tour asked, "Mr. Adams, everywhere I go you seem to be there as well. How do you get to all these tournaments?"

Dad knew perfectly well what he meant. How could a black man possibly afford to take his little girl to all these tournaments? In fact, we mostly drove to our destination, staying a few miles out of town at the Red Roof Inn to save money. Most players and their families flew down, then checked into the Holiday Inn in town. But it was none of this man's business. My father just smiled and replied, "Well, sometimes I drive, and sometimes I fly."

Dad got a kick out of the man's reaction.

"His face turned so many shades of red!"

However small our travel budget, Dad made it work. He didn't need some fancy hotel bar to retire to at the end of one of my matches, for example. Whenever we went on the road, he stocked the trunk of his Cadillac with his favorite bourbon, plus vodka, and mixers for anyone who wanted to join him for his tennis tailgate—usually Jim Gibson, father of Gail, one of my junior doubles partners. "Well James," Jim would say to my father as the sun started to set, "I think it's about time for our nightcap." Then the two men would walk out to the parking lot to have their cocktail.

While we were an unusual sight at the country clubs when I was coming up, today most of our top-ranked Grand Slam tournament champions are individuals of color. Some of the most exciting players of the 2019 US Open were Serena Williams (obviously), Naomi Osaka, Coco Gauff, Madison Keys, Taylor Townsend, Sloane Stephens, and Venus (of course). Coco-mania swept the world at Wimbledon earlier that year, but the fifteen-year-old superstar was far from alone. The press was calling this demographic's dominance in the sport "black girl magic."

Venus and Serena sparked the trend. They transcended our sport, creating a whole generation of young girls wanting to grow up to be like them. The next wave of Americans is Sloane, Madison, and Taylor. Coco is going to be the leader when the next generation of players are coming up. Today's tennis has a healthy pipeline of incredibly talented African American women who are destined to become the next megabrands.

The math shows how far women of color have come. In 1999, four out of the fifteen Americans in the top 100 of the Women's Tennis Association (WTA) were African American. The current sixteen American players in the top 100 include six African Americans, four Latina players, five Chinese competitors, and Naomi Osaka, who is of Haitian and Japanese descent.

That doesn't mean we've come as far as we need to go. There hasn't been an African American male Grand Slam tournament winner since Arthur Ashe won Wimbledon in 1975. France's Yannick Noah, who had a French mother and a Cameroonian father, won the French Open in 1983. MaliVai Washington also reached the 1994 Wimbledon Finals before losing to Richard Krajicek. Today we have Frances Tiafoe, who reached a career-high ranking of 29 in the world before he was twenty-one years old. He could be that next superstar. There are others, of course, but all the young boys in our juniors programs are saying they want to be like him, and that's important, because each community needs a role model. They need to be able to see themselves in someone like Frances to know what's possible.

That's what's been missing for our Latino American players. There are plenty of Latin Americans out there from other countries, including Juan Martín del Potro and Diego Schwartzman from Argentina, and thousands of Argentinian Americans idolize them. But what about some homegrown Latino tennis stars? We need more players to influence the next generation. Although there is Puerto Rican Monica Puig, 2016 Olympic gold medalist, the next player is Ernesto Escobedo who has been ranked as high as 67. But he still has a way to go to be a top player. There is some major talent in our juniors programs, especially from South Florida and Southern California, but their families see tennis as more of a pathway to a college education they couldn't otherwise afford. Hall of Famer Charlie Pasarell from Puerto Rico is the last Latino North American champion.

To rise to the professional level, these underrepresented groups need more opportunity and resources. They can't easily travel the world to play without a sponsor because every dollar they make is spent on travel, training, equipment, or coaching expenses. If they don't have someone who can support them throughout year one, I don't care what color, nationality, or ethnicity they are, they won't be able to break through. At the pro level, pure talent will only get you so far.

Before, during, and after my tenure, the USTA has been trying to bridge the gap, particularly for the top players early on in their development in hopes that they could find their way. We provided resources to multiple communities, including in-school and after-school programming to make sure that we could serve as many people as possible. All the projects we completed at USTA Billie Jean King National Tennis Center through our strategic transformation, from the rebuilding of our grounds to the development of the USTA National Campus in Florida, which is open to the public, has been with the intention of making everyone feel welcome in our sport.

The USTA's head of player development, Martin Blackman, an African American who earned a tennis scholarship to Stanford, has made it his mission to make tennis more appealing and affordable to African Americans and other minority groups. One way for these players to get to the professional level could be through college, but the competition for those tennis scholarships is fierce. As always, it's the families with money who can afford to pay for private lessons for their children who have the edge.

That's why Martin has put so much energy into our Excellence Team program, which includes a focus on education, through the USTA Foundation. The USTA is also developing more community-based African American coaches throughout the country who can bring free lessons and mentoring to kids at the local clubs who may have the talent but not the financial wherewithal to realize their full potential—much like the support I received when I was coming up in the sport.

Our goal has always been to make tennis look like America. Whatever our demographic, we want to make sure that our sport mirrors that. I'd just joined the USTA board, in 2005, when we appointed our first chief of diversity and inclusion. Back then the department was one person, but she worked hard to raise awareness. It took time to learn what was needed, how to implement it, or even to properly define diversity and inclusion.

Our four pillars of diversity are African Americans, Asian Americans, Latinos, and the LGBTQ community. It makes good economic sense to be inclusive of these and other minority groups, because they have tremendous spending power. Not all minority players are necessarily poor and underserved. We have millions of professionals with means. Their kids are playing because they love the sport, not because they are coming from some inner-city program. According to the British Lawn Tennis Association, it costs more than $300,000 in travel, tournament fees, coaching, and equipment to raise a champion from the ages of five to eighteen, and it's the parents, not just sponsors and investors, who are paying for it.

We have the means. Latino Americans alone have $1.3 trillion in spending power, and African Americans are right behind them. We buy apparel, sneakers, and rackets. We pay the dues for private tennis clubs. The population is shifting in terms of size, spending power, and upward mobility, so having representatives who reflect these consumers is good business.

We also need diversity beyond what people see on the court. In order to move the dial toward inclusion and bring about authentic change, the USTA and other organizations in the tennis world need to reflect internally what they'd like to see happen externally. Diversity and inclusion are not just a matter of optics. If your behind-the-scenes team doesn't accurately reflect what the world of consumers looks like, it comes across in the myriad decisions that get made. From the Grand Slam tournament greeters to the coach giving lessons at a USTA Juniors program in Houston, Texas, there's a certain comfort level when fans and participants of the sport can see people who look like them.

Appointing me as president, CEO, and chair in 2015 was an obvious positive step in the right direction. I represented a huge change to what was deemed to be a country club culture of the USTA. I'm waiting for the day they put my face up on that esteemed portrait gallery. Though I'm not sure why it hasn't gone up to this day in 2020. It makes me wonder do they not want to see a different face alongside all the others who are similar in hue. As gratifying as it will be, I am mindful of the absurdly long time for Althea Gibson's statue to go up on the grounds at the start of the 2019 US Open, which happened to be the same time Arthur Ashe was finally inducted into the Tennis Industry Hall of Fame. But hey, we got there eventually!

Don't even get me started on how unlevel the playing field has been for women. Tennis players are the highest paid female athletes in professional sports. The US Open was already well ahead of the other Grand Slam tournaments, introducing pay parity in 1973 compared with 2007 at Wimbledon, and the 2019 US Open champion's prize money was $3.85 million, the largest purse in the history of the sport. But it wasn't until players like the Williams sisters, Anna Kournikova, and Maria Sharapova came along that off-court earnings for female players began to exceed the $10 million mark. That said, Serena Williams, who made $18.1 million in 2018, falls far behind Roger Federer, who earned $77.2 million during the same tennis season.

As head of the USTA, I did my utmost to remove any disparities, including recommending changes to the seeding rules for women returning to the game after maternity leave. Seeding is a system that we use throughout tennis to separate the top players in a draw, with the top seeded players being those whose rankings have earned them that position. In Grand Slam tournaments, there are 128 draws, and these tournaments in particular have the right to seed players based on their results over the years that might place them above someone ranked higher than them. Being seeded means you don't have to play any of the top 32 ranked players in the first three rounds of the tournament.

But that wasn't the case for Serena, who suffered a huge disadvantage when the French Open refused to seed her in 2018. At Roland-Garros, her first Grand Slam tournament since giving birth, Serena had to face top players in the early rounds. After playing only six matches all year before arriving at Roland-Garros, she played six matches in a week, while also playing doubles with her sister Venus. The excessive match play resulted in an injury, which caused her to withdraw in the fourth round, where she was to meet one of her biggest rivals, Maria Sharapova. Had she been seeded, Serena may well have had the same results or gone a little further.

In an office environment, it might look like a woman returning from maternity leave to find she has been demoted. One of the greatest, if not *the* greatest players of all time, was not being protected. Serena was ranked 453rd after losing her points and ranking due to her inactivity for over a year. It essentially forced Serena to prove herself all over again. Although she wasn't the player we knew before her pregnancy took her from the tour, she fought hard and did well.

This incident reignited the conversation we'd already been having about protecting our best players and ultimately the tournament, because champions like Serena are our ticket sellers. We tried raising the issue a couple of years earlier, when Belarusian champion Victória Azárenka was returning to the tour after giving birth, but Serena's situation gave us new momentum. We decided to announce our plan to seed Serena ahead of Wimbledon, which comes before the US Open, in the hopes of pressuring officials of the All England Lawn Tennis and Croquet Club to do the same. They did, granting her a seed of No. 11. I like to think we played a role in pushing their hand. That move helped Serena to advance to the Wimbledon Finals, proving our point.

Helping new mothers keep their status in the rankings, which is key to qualifying and playing in the more lucrative tournaments, was a huge step. But it still hurts my heart to see the inequities that persist for women in other sports, including women's soccer. Pioneers like Billie Jean King continue to push hard for pay parity and media exposure for all professional female athletes, who, to this day, still only receive 4 percent of national media sports coverage.

At the International Tennis Federation, we launched a gender equality media campaign called "Advantage All," which highlights the point that tennis is an equal advantage sport. Our theme centers around making sure that tennis speaks with one voice for the male and female game. In tennis the term is "deuce": the tie score of 40-all in a game, when both sides have an equal opportunity to win. When both players are strong and the match reaches a point where it could tip in anyone's favor, it's the most exciting tennis to watch. That's what diversity and gender equality can do in and outside of tennis.

The USTA has been among the most proactive of sports organizations about modernizing its culture, having implemented several new policies to accommodate players in addition to evening the playing field for male and female players, including feature court scheduling and enhanced broadcast scheduling. Putting them on the main courts has brought the players and the WTA more exposure. But living and breathing it in the day-to-day running of operations is another matter.

That's why in 2012 we hired D. A. Abrams as our chief diversity and inclusion officer (CDIO). He left the USTA in March 2019. But during his tenure, Diversity and Inclusion grew from a four-person team to six people, with D. A. reporting directly to the CEO. Unlike some other organizations, we've always had the CDIO report to the top rather than to the HR director, because it ensures that diversity becomes a way of thinking and being that links across all functions of the organization. A core value of the USTA, diversity cannot be just some adjunct of a department that allows the organization to tick off a checklist of minimal requirements. It also gets board members more directly involved in various diversity and inclusion initiatives, reflecting how much of a priority diversity and inclusion have become.

I've known D. A. since we both played in tournaments for the American Tennis Association, which was founded in 1916 when there were strict laws preventing African Americans from playing in USTA-sanctioned games. I got to know him better when he was running the USTA's National Junior Tennis and Learning (NJTL) program and leading our community outreach. When the CDIO position became vacant I was on the search committee, and choosing D. A. to fill the slot was a no-brainer because of his experience, which included his roles as executive director for two different sections: Missouri Valley and later Eastern. D. A. has also written multiple books as a D&I strategist. He came up in tennis as a junior player in the 1970s through the National Junior Tennis League in Philadelphia, so he knew firsthand the challenges of accessing the sport as a young African American kid from the inner city. But it was his strategic and thoughtful approach to diversity and inclusion that impressed me most.

Our diversity and inclusion policies are based on the understanding that most people don't intentionally set out to exclude others. The root of the problem is unconscious bias—something that we, as humans, have all been susceptible to at some point in our lives. It's why, twice a year, our team travels across the country doing training for cross-cultural dexterity. That basically means our board members and staff benefit from a series of nonconfrontational, nonjudgmental workshops that help us get to the root of where this bias comes from, because you can't have meaningful change without self-awareness.

The USTA D&I team led an unconscious bias session for our new administration's board meeting in Las Vegas. One of the exercises was called "Insider/Outsider," consisting of a group of ten USTA representatives sitting at a table. One of them was randomly chosen to step outside the room, with the instruction that they had to try to get back into the conversation as quickly as possible. Those inside the room were told under no circumstances to let the person who'd been standing outside break into their conversation. When the outsiders in each exercise returned, they spent the next five minutes trying to infiltrate the conversation and get back into the mix to no avail.

At the end of the session, one of the D&I staff members sat down with the group to help them unpack what had just happened.

"How did that make you feel?" she asked the rep who'd been left outside.

"Not so good!" he admitted.

"So how do you think this affects the work we do on committees or in the workplace?" she asked, driving home the point that failure to be inclusive hurts not just the person being shut out of the conversation but also the team, because it deprives them of that individual's potentially valuable insights.

These workshops demonstrate how we might be more predisposed to what is most familiar. When we see individuals who don't look like us, for example, we might just assume they're not as capable. On the flip side, when I walk into a room full of men, which has often been the case in my career as a global representative of my sport, the first thing I notice is another woman. I make a beeline to say, "Hey, how are you doing? I'm Katrina." It's the same in any culture. If I'm a person of color in a room full of Caucasians, I'm going to head straight over to the people who look like me. If you are Caucasian in a room full of Latinos or another culture, you also beeline. It's only natural when, for so many years, there was no one else who looked like me. Ironically, when I started playing tennis, I didn't know that it was a predominantly white sport, even though I should have recognized by what I saw on television. I started in a black program. I had a black coach. I went to another black program indoors that winter, followed by a black program the following summer. My first tournament was the ATA Nationals, a black event. So I didn't know anything else until I started playing in USTA sanctioned events and was no longer in a fully welcoming environment. But it gave me motivation to kick some butt.

When I was a junior tennis player growing up in Chicago, many tournaments I played were on the grounds of the country clubs dotted throughout the suburbs and small towns of the Tri-City area. It's not that they were segregated but this was during the late '70s and early '80s. You didn't see other black kids mingling in these crowds, much less sweeping through each match and taking every trophy. I was the only one in my age group.

So, my response was to walk in like I owned the place. I told myself I belonged there, which I did, and my body language left no room for interpretation. I'd enter and make a beeline for the crowd of other junior players I'd gotten to know as we practiced together and competed in junior district, state, and section championships. No one ever stopped me to ask what I was doing there or check my credentials. I wasn't going to give them the chance to embarrass themselves.

I've always been told I have a certain strut. I move along briskly with my back straight, my head held high, and my eyes looking ahead to where I'm going. I may put a polite smile on my face, but I don't slow down or wait for permission to enter. I never hesitate. Friends and colleagues who attend Grand Slam tournaments around the world with me know they must keep up. They even joke about having to train hard to build their endurance so they can stay in my wake. If they even think about stopping to fumble around for their VIP badge, I'll be long gone.

That sense of owning the arena isn't entitlement. It's not arrogance and it doesn't announce itself with brashness. It's more like a quiet confidence. I've always had a silent yet unwavering belief in myself as a leader. But I'm well aware that it's not the way everyone sees me. Intentionally or not, people judge me differently as a woman of color. I must hold myself to a much higher standard than someone else in my position, never allowing myself to relax or let my guard down while I am in the public eye.

I can sometimes present as guarded to prevent people from

coming at me, being microaggressive or beyond. Unfortunately, in America most white people's rare exposure to black people is through news programs, not on a regular personal level. As a result, they find it difficult to believe there are competent, intelligent, articulate, and visionary people of color—because they are not in our arena. And when they come across us, they may make improper comments. For instance, I had just given a speech and someone came up to me afterward and said, "What a great speech. You were so articulate." I did not perceive this as a compliment nor do other black people. Why wouldn't I be able to express my views clearly? Why wouldn't I give a thoughtful and informative delivery of my views? Is this a statement you make to your white friends? I think not.

When the topic of race comes up in conversations, to diffuse the topic my white counterparts would often say, "I don't see color" or "I don't see you as black." Really? How can you not when that's what I am. Those comments actually hurt more than they make me feel embraced. Why? Because it means that you don't see me for who I am. You only see me for what's convenient for you.

I had become accustomed to shoving racially insensitive comments under the rug but not anymore. It's important that we educate one another and hold each other to a higher standard of humanity. Let's eliminate rooms where there are "onlys" and treat those who are there with respect. People of color should not have to apologize or downplay their presence and contributions to make others comfortable.

I also used to downplay the significance of being a "first" or an "only" in order to fit in and not bring too much attention to myself. But now I know it is important to own one's identity, as it creates your next opportunity. We should scream it loud and be proud, as we are walking the path to accomplishment that our ancestors laid.

Unconscious bias is no less insidious than overt racism because the person acting on it doesn't know better. The only way to move past it is to be more intentional in the way we look at people, whether that's through the hiring process or simple human interaction. Think of unconscious bias as a glacier. All we see is the piece of it that sticks out of the water, not the larger portion of it that sits below the surface. You're assessing someone on the first look before even having a conversation. When you judge only what you see on the surface and never allow yourself to look at all the other aspects of this glacier, you're missing out on so much. That's why when you're looking to hire someone, it's important to see beneath the surface. Look at the substance of the person, their knowledge and experience. Don't be afraid to make a different decision because you've always done it a certain way. We need to judge people on who they truly are.

In that process, make sure you have an array of different people—men, women, black, white, Latino, Asian, older, younger, LGBTQ—then choose the right person from that group based on their skill set. It may not be a person of color or a female, but as long as you know you have gone through that process you can feel comfortable with the decision you make. If it happens to be a white male again, that's okay. You might find someone else in that group would be perfect for another role within your organization or business. People come in with multiple skill sets, so even when their expertise doesn't fit neatly inside a box, they can add value. Take a risk on someone with that level of potential. In the right role, it may only take a little upskilling to bring them up to par.

All the work we've done on diversity and inclusion has had a powerful impact on members of the USTA team, be they coaches, National Junior Tennis and Learning Network chapter leaders, program leaders, or the administrative staff of volunteers. Everyone has become much more intentional in their decision making. We even initiated a "blind" approach to our hiring process. We aren't perfect and have a long way to go, but as long as diversity and inclusion is embedded into the daily business operations, the USTA has a chance to be the true leader in sports and business.

Before a job candidate's résumé comes up for consideration by a potential boss or department, their identifying information—names, colleges, where they live—is removed from the document. The only items left are experience, training, and skill sets. Then we assemble a diverse interview panel. Once the candidates are selected based on these qualifications and invited to meet with us, you'd be surprised how many more highly qualified candidates from the four pillars show up. Without that initial filter of unintentional, unconscious bias, they get a fair shot, and the USTA gets the benefit of a much larger selection of well-rounded job candidates who also happen to match the markets we want to penetrate.

This is about taking a chance on someone with a different mindset who can help broaden your perspective and contribute to your own growth as a leader and team member. We *all* need to break out of certain habits of mind.

Just before I officially took up the position as head of the USTA, I was invited to join a select group of industry and community leaders to experience the annual Joint Civilian Orientation Conference (JCOC), where each branch of the military invites senior industry leaders to experience life in the military firsthand.

I first became aware of JCOC through one of our volunteers,

Robin Jones, with whom I worked closely to strengthen our relationship with the military through the USTA military initiative. She and Tim Richardson were the consultants who created the Adopt-a-Unit program, which sends care packages, among other things, to our servicemen and servicewomen who are deployed. In addition to the usual toiletries and snacks, we packed portable tennis nets, rackets, and foam balls that they could set up anywhere and play tennis. The intention was to keep them connected and provide a few moments of fun whenever they had some downtime. I was so engaged and moved by this initiative and the ways it could elevate our cause that I stepped up to be a spokesperson whenever I could.

The JCOC was established after World War II to give civilians an understanding of the capabilities of service members so that they can be considered for future employment opportunities. It's an immersive, six-day experience, beginning with three days of briefings at the Pentagon, then flying off to various military installations across the US.

What an adventure! We got to fire weapons with the Green Berets and the US Marines. We rode in tanks, sailed on US Coast Guard cutters and navy aircraft carriers, and flew in military helicopters. It was like being sent to the most exciting sleepaway camp for adults in the world. But it wasn't all fun and chopper rides.

We arrived at Marine Corps Base Camp Pendleton at 5:00 a.m. The hosting officer greeted us warmly, explained the core values of the marines, then thanked us for supporting the military and being part of the journey.

"Your activities today will take you through what it's like to be a recruit," he calmly informed us.

As we were about to depart the bus, the drill sergeant asked if we had any final questions, then politely told us to stand on the yellow-painted footprints and put our water bottles down in front of us.

"Any last questions?" he asked.

We all shook our heads, and the second we did so, he dropped the soft tone and started screaming: "In position, in position! Stand up straight! Shoulders back!"

For the next ten minutes, we got yelled at. He treated us exactly the way he would have treated brand-new recruits. We got taken to a room where new marines were told to call loved ones to let them know they arrived safely and warn that they wouldn't be hearing from them for the next few weeks. We read an oath on the wall, then we were marched down the street.

"Left, left, left, right, left, forward march!" the drill sergeant screamed as we were led to our barracks and instructed on how to make proper beds (nice and tight with precise folds at the corners).

The point was to have us experience how those first few moments must feel to an eighteen-, nineteen-, or twenty-year-old when they first embark on this path. Everyone's scared and wondering what the heck they've gotten themselves into. But those who stick with it say it's the best decision they ever made.

By far the best part of the whole experience was during lunch in the mess, speaking to the servicemen and servicewomen of all ranks. After a couple of tours, they become experts in their craft, with top-notch training as engineers, statisticians, logistics analysts, mechanics, you name it. While many members of the military are college educated, some didn't complete high school or have troubled pasts with conviction records and can't get jobs. These individuals are the best at their craft, with a proven ability to conduct themselves with grace under pressure. They've risked their lives in both real combat and simulated situations that can be just as dangerous. Living by a code of honor and discipline, they've blossomed and matured as human beings and professionals. They've risen to the occasion when put in critical leadership and team positions where their conduct and decision making can affect the safety and well-being of an entire fleet. They made me proud to be an American. We made a point of employing veterans, yet few other organizations will hire them.

That's why we need to view each other through a whole new lens. The less inclusive we are, the less welcoming we are to people with different experiences and backgrounds, the more we'll lose. What will move the needle forward in making the case for inclusion is an understanding that diversity isn't just an ideal to aspire to. It is essential to the creation of a team of excellence that yields tangible business results. It's the only way to truly win.

5

A GOOD LOSS

For me losing at tennis isn't failure, it's research.

—BILLIE JEAN KING

We were at a resort in Maui, hosting Germany for a women's international team competition called the Fed Cup, the World Cup of Women's Tennis, which is the women's equivalent of the Davis Cup, both ITF competitions. The teams representing various countries compete through a series of first-round ties, with matches played on a home-and-away or knockout basis, usually over a weekend. The USTA was responsible for staging the 2017 event, with all the flags, pageantry, and ceremony that goes with it. As part of the protocol, we'd hired a local opera singer to perform the German national anthem. The guy was a professional, supposedly, and although we'd usually do a run-through with both teams present, he was someone the local entertainment team had worked with in the past, so we figured it would be fine.

On the opening day of the match he started singing the first verse from "Deutschlandlied."

Deutschland, Deutschland über alles, über alles in der Welt.
(Germany, Germany, above all, above all in the world.)

It was an outdated version of the anthem that the Nazis had used for propaganda!

We first realized the gaffe when we saw the stricken faces of the German fans and players, who locked arms, turned their backs, and attempted to sing the correct lyrics in an effort to drown out this unprepared Hawaiian tenor. I exchanged looks with Stacey Allaster, chief executive of Professional Tennis for the USTA, who oversees all professional tennis properties, including team events, such as Fed Cup and Davis Cup. We were simultaneously confused, as was the president of the German Tennis Federation, Ulrich Klaus, who was standing to my right, looking uncomfortable and turning red.

Ulrich was flanked by the ITF representative, Ingrid Löfdahl Bentzer, who also spoke German. When they started chatting under their breath, I knew something was wrong. We left the court and scrambled to figure out exactly what had happened. When we did, of course, we immediately issued an apology. After apologizing to Ulrich, I immediately approached the German team captain, Barbara Rittner, a fellow competitor from the WTA Tour, to apologize. She was completely distraught and angry and wasn't having it and I totally understood.

"In no way did we mean any disrespect. This mistake will not occur again, and the correct anthem will be performed for the remainder of this first-round tie."

The next twenty-four hours were intense. We apologized numerous times, publicly and person to person. But the Germans were still livid, understandably so.

"I thought it was the epitome of ignorance," said Andrea

Petkovic, one of the players, after losing a match to US team member Alison Riske. "I've never felt more disrespected in my whole life, let alone in Fed Cup, and I've played Fed Cup for thirteen years now and it is the worst thing that has ever happened to me."

Barbara called what happened "an absolute scandal, a disrespectful incident, and inexcusable."

As chair of the ITF Fed Cup Committee, I needed answers. But my role here was USTA president and host to our visiting team. It can be tough wearing two hats at a moment of crisis. I don't believe in micromanaging. It's my way to trust our team of experts to handle these details or bring me in if they have questions. But I made the mistake of assuming we had this. Our team is well seasoned and made up of individuals who normally cross all the t's and dot all the i's. But there was no German-speaking person at the rehearsal. As a result, we deservedly became fodder for news outlets around the world. That an anthem so closely linked to the horror of the past should resurface at an event intended to celebrate a nation's finest athletes was an epic fail. But we learned from it. Oh boy, did we learn. Everyone talks about learning from failure, and that was certainly the case this time. Never again would we skip a step in our system of checks and balances.

However embarrassing or devastating the loss, my training as a tennis player has taught me how to absorb the lesson and move on quickly. Tennis is a preparatory sport for becoming who you are. It builds character and resilience both inside and outside the sport. It teaches us what it means to have a good loss that helps us to develop smarter strategies for the next match, along with the endurance and persistence that come from repeatedly falling on your face.

In this sport the majority lose most of the time, especially as we climb up the ranks, unless you're the top player, because the competition of great players is that much more intense. Consistent, years-long winning streaks are extremely rare in tennis. Everyone is vulnerable. If you look at the stats of even the world's highest-ranked players, you're going to see quite a number of losses next to wins, because out of the hundreds of tournaments they play, there can only be one victor. An exception is Jimmy Connors, the player with the most wins in the history of the Association of Tennis Professionals with 1,274 over twenty-five years. That's a win-loss ratio of more than 80 percent. But he *still* managed to lose 282 matches. A woman, Steffi Graf, holds an even more impressive ratio, having won 89 percent of her matches, with an overall singles record of 900 to 115.

But even those two experienced losses. It's the nature of the game. To develop into a great long-term champion, enduring those losses is a good thing. Oftentimes players who peak too early don't go all the way because they didn't lose enough. You must understand how you lost in order to win. The only way to become a Federer, for example, is to learn to deal with adversity. You are either going to come off the court a winner or a loser of the match, but not before losing plenty of points, games, and sets. There is just no avoiding it. On the pro tour, where we are competing with hundreds of other hard-driving players, defeat is our constant companion.

It's always a possibility that an up-and-comer with a low ranking could upset a top-seeded player. Every year, someone younger comes along to cause an upset in a match against a reigning champion, like Naomi Osaka, Coco Gauff, and Bianca Andreescu—players in their teens, and barely out of their teens, who've got the energy, strength, and endurance of younger limbs combined with state-of-the-art training. That's what makes the game so exciting. In any given match, it's possible for kings and queens of the courts to get dethroned by some upstart player from out of nowhere.

Imagine losing in front of thousands of people, day after day during a season? Imagine looking up at the scoreboard, where it says 6-0, 3-0, knowing darn well it's a disaster, yet maintaining your composure as you fight to the end? It's devastating. But enduring those defeats with grace is part of what makes a champion. That's how I felt in 1993 in the third round of the Australian Open, where I got waxed 6-0, 6-0 by a young, brash, strong champion, Jennifer Capriati. It was the farthest I had gone in Grand Slam tournament singles play since I had lost to Chris Evert.

I was as confident as I had ever been going into the match. I was playing well, feeling great about my fitness and my performance in the previous two matches. Yes, Jennifer was young, powerful, skillful, and on her way to becoming No. 1 in the world, but I felt that I could win. I thought my serve-and-volley game was the perfect weapon to put pressure on her and rush her shots into errors. Boy was I wrong! The sooner I finished my serving motion to move forward, the quicker the ball was whizzing by me for a winner. I totally underestimated her strength, accuracy, and speed. She was fast, and when she arrived to the ball, she hit it with marked power and precision.

The match was over before it started. My heart and mind were racing from point to point and I couldn't slow down for the life of me, even on the changeovers. I allowed Jennifer to dictate the pace of play, which suited her game extremely well. I was outmatched physically and mentally. This was a lesson that I would never forget, with Jennifer using me as her student and putting a clinic on Court 1, the second-largest court at Melbourne Park. It was especially demoralizing to be finished off before I even got started, but I never allowed that result to completely deter me from working harder. This situation can happen in any discipline, where you don't slow down and acknowledge how the tone and tide of a meeting may be wavering. Own the feeling and make adjustments to turn the situation around.

The pain of defeat builds character and resilience. Some of the greatest players of all time (GOATs) have had the strongest runs of their careers after coming back from an injury or enduring a string of defeats. GOATs *hate* to lose. That feeling stays with them so long that they never want to lose again. Losses keep them moving forward by fueling their drive to overcome and win. As the legendary Peachy Kellmeyer, the first director of the Virginia Slims circuit of the WTA Tour, once told me, "Tennis teaches you to get up the next day and start all over again."

It's about the art of losing well. Roger Federer is a great example of a champion who has learned how. A cool and classy customer on and off the court, he used to simmer over his defeats for days, admitting he often cried after losing every match. At Roland-Garros in 2019, Roger launched a ball into the stands because he was so frustrated about losing to Rafael Nadal. But he doesn't hang on to that feeling. Instead, he lets off steam and quickly moves on. After losing to Novak Djokovic in straight sets during the 2015 Indian Wells Final, he said he planned to move on from the match "in like twenty-five minutes or so."

I hated losing. Hate, hate, hated it! As a kid I was as guilty as anyone of my share of hissy fits on the court. I earned the nickname "Jane McEnroe" from throwing down and breaking countless rackets after losing a game point or disagreeing with a line call. I had no interest in the runner-up plate; I wanted the "big" trophy. Coming in second was completely unacceptable to me. This behavior was completely unacceptable to my parents, as it embarrassed them tremendously. As one of the only blacks playing, I needed to be more mindful of how I conducted myself at all times.

Thankfully, I grew out of it. Well, kind of. I also came to the realization that you are more likely to lose when you have a worthy adversary. In my college and early years as a pro, my archrival was Stanford player Patty Fendick. Her college team won the NCAA title three times. She won two NCAA singles titles and had a 57-match winning streak. In 1987, she was named ITA Player of the Year. We battled each other in singles and doubles matches throughout our careers. As my fiercest competitor, she brought out the best in me and I brought out the best in her, although I always got the edge.

Pitting yourself against someone strong makes you grow. That's why the USTA changed the competition format for entry-level tournaments. In the past, families would drive three hours to watch their kids play one match. We have since created a different format so that these juniors can participate in two or three matches minimum, with short-scoring formats. The kids play for an hour, take a break, then go out and play for another hour. This structure gives young players the experience they need to excel in their sport, putting the focus more on development, not just winning or losing.

They also learn to overcome their fear of losing, which can be debilitating for even the most talented players. In fact, fear of losing can stop you from winning. Steven Wilson, one of my graduates from the Harlem Junior Tennis and Education Program (HJTEP), was a brilliant junior player. In each age group he ranked in the top twenty of New York City. He kept making the final rounds of tournaments across the metropolitan area. But, as good as he was, he kept getting tight in the finals. His brilliance on the practice courts wasn't translating into the big wins. The moment he heard the word *tournament* he was shaken and started to stress over the potential outcomes.

"Steven, even the best of the best sometimes get nervous and that's okay," I told him one day. "But you beat everyone on the practice courts because you know you're good at this and you're just out there having fun. Try to bring a little of that nonchalance to your tournament playing."

It's not that I wanted him to get comfortable or complacent about the possibility of losing. That's going too far. But you must give yourself permission to experience loss. He was so fearful of making a mistake it messed with his head and he couldn't be in the moment.

Steven started winning a few more finals when he changed the conversation he was having with himself. Instead of dreading the possibility of losing a point, a set, or a match, he allowed himself to experience the inevitable setbacks and move on. Later, at Wilkes University, a Division III school, he made Player of the Year twice and got into the nationals four years running. But he never could fully overcome his finals jitters. If he had done so, I believe he could have gone to a Division I university.

Of course, the best way to get over losing is to win. It builds confidence and reminds you that the setbacks are temporary, and your moment will come. Justin Holmes, one of my kids from the HJTEP, will never forget that feeling of winning his first tournament at nine years old. All his coaches and family members were in the stands cheering for him. There were hugs all around when he came off the court. Justin loved that feeling you get when you know you've made the people who've trained you and cared for you proud. He admits that at that age, "I wasn't a terrible player, but I wasn't good." But that victory showed him what was possible and inspired him to get better.

"To have the people I loved and respected most look at me like I was the winner," he told me, "I wouldn't trade that for anything."

But nothing develops backbone more than losing. It's a reminder to be humble and empathize with what others are going through. In fact, it can be the game's biggest teacher, helping you stretch and grow. We don't really learn from wins because we tend to assume that we did everything perfectly. When you lose you must reassess yourself on every level. It's the tennis court that brings out your true personality.

The risk of losing is also part of the fun. It makes the win that much more meaningful. It's also a way of testing your own strength and figuring out how to adjust your strategy. There's no other sport that provides the same level of competition with yourself. It's not just about playing against someone else but challenging yourself to be the best. The game teaches you to deal with constant adversity, helping you to build up your self-esteem and your physical strength as a result.

The best loss I ever had was that fourth-round Wimbledon match against Chris Evert, because it forced me to take a good hard look at what I was doing both on and off the court. I looked at my pregame strategy, my mental game, as well as my general physical fitness. The way I treated my body was a huge lesson. I learned that the way to a win wasn't just about exploding with power right from the start of a match; I had to be able to endure from start to finish. In the tortoise and the hare race, I was the hare who risked burning out. I needed to take steps to become more of an Energizer Bunny.

My coach, Willis Thomas, had been pleading with me to eat better. My first year of the professional tour I was enjoying the baguettes and cheese of Europe a little too much. I didn't understand why other players on the tour were weight training and eating salads when they could be relaxing and enjoying a nice meal on their off days. Nothing tasted quite as good as an ice-cold beer after a practice on a hot summer day. I figured I was young and strong enough to burn off the extra calories while I was playing. I was wrong, and the loss against Chrissie proved it.

To receive those lessons, you also need to learn to shake it off. I have friends who remember every single bad thing that happens to them. They can't forgive people. It's as if they keep a mental accounts ledger of all their past grievances and disappointments. Take that time to grieve or vent to a friend or a therapist. Release it. Have a good cry to wash out your soul. But don't rinse and repeat. Telling yourself about your disappointments over and over again will only reinforce a bad attitude. That doesn't mean you forget, but don't let it haunt you. Instead, step up and take responsibility to improve the situation. Fix it!

You must face down problems at a personal and professional level, then set them aside so that you can enter the court with a clear conscience. If you don't, though you may not realize it, you are exerting internal energy, which leeches off your physical endurance. You feel heavy, and it shows up in your playing—all that moping is visible to others. During a televised match you'll hear sport commentators say, "Wow, something must be going on in that player's life!" But champions learn to overcome the losses, no matter how profound: financial challenges, a breakup with a spouse, an illness in the family, a bad report card at school, or a demotion at work. Whatever is weighing them down gets set aside the moment they step onto the court.

This plays out in the business arena as well. Losing a $100,000 a year donor to the HJTEP as a result of the financial crisis could have been devastating to our budget, which relies heavily on these funds. Most of our programs are paid for by smaller, individual donations, so that's a large chunk of money to replace. It involves reallocating expenses, making up for lost income without taking away any benefits to our players. But the worst thing you can do in a situation like this is panic. Instead, we look at the situation as not just a temporary setback but a challenge to reevaluate our spending and an opportunity to attract a bigger and better donor. To look at a circumstance like that any other way would have been detrimental to our progress. If we couldn't view things in a positive light, potential new donors would have sensed our desperation in meetings and been turned off.

When I was playing professionally and tournaments were my business, resilience was the difference between winning a purse and losing one. In August 1991, I was playing in the Virginia Slims of Albuquerque in New Mexico, having my best performance in many tournaments, when I received the news that my maternal grandfather had died. At that point I had reached the quarterfinals, to play Julie Halard from France, and I was still in the doubles, partnering with Isabelle Demongeot, another French player. I had to make a big decision. Either stay and miss my grandfather's funeral or go and risk not being prepared for my next match. I'd already missed my paternal grandfather's funeral because I had a college match, and I always regretted it, so I chose to go, traveling from Albuquerque to Carthage, Mississippi. There were no nonstop flights to Jackson, the closest airport. I arrived just in time for the services and then left early the very next day to return to competition, thanks to the scheduling of my match to be late in the day. Exhausted from the travel, I lost my singles match 2-6, 6-4, 0-6. I was lethargic in the beginning, got it together and played well by the second set, then ran out of gas, physically and emotionally. It was a match I felt I could have won but being there for my family was way more important. I'd made my choice, and I didn't regret it even though it led to a loss. I shook it off, reminding myself there would always be another match, which of course there was the next day, after a good night's rest. I went on to win the doubles match the following day and a couple days later I won the title.

The same is true when I perform in an office setting. While chairing a meeting I must focus on what's before me and not whatever losses, failures, or disappointments took place in the other room. Negative energy is a focus killer. I can't adequately tune in to what others are saying or engage in the conversation if whatever happened the night before is still weighing on me. Imagine trying to do that when you've got to get through an agenda of dozens of items with high-powered board members, global representatives of the sport, and donors all looking to you to keep things moving. So, if I find myself drifting, I take a deep breath, drawing in as much air as my lungs will hold, then release it, blowing it out to clear the brain fog and get back to the present.

Great athletes understand this. They know how to master their own disappointment. One of my favorite moments was in Atlanta during the 2001 PGA Championship, when I followed Tiger Woods from the gallery. After he missed a five-foot birdie putt I said, "It's okay, boo." At that point I was a few feet away from him, green side. Even though he was annoyed with himself, he looked up and smiled. He recognized the tone of my voice as that of a black woman and curiously looked up at me. Tiger knew exactly what I meant, or so I like to think he did.

When I was competing, I would tape a small piece of paper onto the throat of my racket with four or five sayings to strengthen my mental game. One of them was "stay in the moment." When I was on the verge of closing out a match, I had to tell myself, *Don't think ahead; don't look back.* Too often when you're on the verge of winning, your head is already in the locker room. Before you know it, you've lost the match. The same goes for closing a business deal before it's signed. If you anticipate too far ahead or berate yourself for missed opportunities, you could overlook important details that could cause the deal to fall apart. Just think how many times you spent millions of dollars from winning the Powerball in your mind. We all get ahead of ourselves.

Patrick Galbraith, my successor as USTA president, recently shared how he got his leadership and business training from facing down obstacles on the tennis courts. Like me, he grew up playing junior tennis and joined the pro tour for twelve years. We played at the same time, though never with each other. (I like to tease him that he never played Mixed Doubles with me, although I did once beat Patrick and his partner Alicia Molik while partnering with Brent Haygarth.) Instead, we became teammates as the first former professional players to become chairman and president of the USTA board. Patrick not only lead the USTA in 2019–2020, but he also worked a full-time job as a wealth manager at the global banking giant UBS. Like many professional players, he became interested in the markets when he had downtime on the tour. He invested his winnings so successfully it led to him managing the investment portfolios of private clients and brought his expertise to the USTA Investment Committee. He served as the treasurer for three terms before shifting to be the first vice president before ascending to the presidency.

What makes Patrick great in both roles is that he doesn't let the fear of losing stop him from making bold decisions.

"Every day is a different day," he told me. "I don't know what's going to happen in markets or which client is going to call with issues, and I love that. It forces me to think on my feet."

He advises his clients not to fear making a mistake "because you are never going to get in or out of an investment at the exact right time." Instead, he devises long-term financial plans that leverage weaknesses and turn them into strengths, so that ultimately the wins outweigh the losses.

When we transformed the USTA Billie Jean King National Tennis Center in Flushing Meadows, we were trying out many things that had never been done before. For a nonprofit organization, we were being extremely bold in terms of financial investment. Our $600-million-plus update was ultimately successful, but at times it was a logistical nightmare, with plenty of mishaps along the way. Attempting something so huge, it was inevitable that we'd bump into obstacles.

One year, a major issue was the state of a few of the tennis courts. Water and moisture had gotten trapped under the surface, causing it to bubble up. The court surface lost adhesion and pieces of it were coming off. Imagine that happening in Arthur Ashe Stadium in the middle of a Grand Slam tournament. We installed the new court surface with lasers to make sure everything was smooth, but some spots had not been identified, resulting in bad bounces. Fortunately, the bubbling was noticed and tended to immediately, and we beat a bullet by discovering this pretournament. We took the affected courts out of commission during the qualifying rounds, stripping them down with a concrete grinder and resurfacing them from scratch. It was challenging not having these courts online in the early days of the event, but we were able to manage and reschedule practice matches on other courts. In the grand scheme of things, this was a minor issue for an enormous construction project—but the week before the US Open, this was major.

Professional players notice the slightest deviation, so for the sake of our brand we had to get it right. Before another US Open, during practice week, Rafa had been practicing in Arthur Ashe Stadium when he noticed the ball deviating from its trajectory. A recreational player may not have noticed it, and there was every chance we might have missed it. We brought in one of our engineers, who noticed a tiny nodule in the resurfacing. We took the court off-line for two days, fixing it just in time. Since then we've changed the material composition of our courts' surfaces, and thankfully have not experienced that kind of havoc again.

Another unforeseen US Open hiccup was easier to resolve. In the middle of a match, at the very moment the players needed silence, someone from the food concessions decided to dump glass bottles into a nearby recycling dumpster, creating loud shattering noises. Until the roof was installed, these sounds were barely heard. But now, everything is amplified. We immediately called the operations team to schedule the glass dumps between sets or matches. It might seem obvious, but when you have twenty-two thousand people from different vendors dealing with overflowing trash as they service hundreds of thousands of spectators, all anyone is thinking about is getting their individual jobs done.

Trial and error can get you to a place of near perfection. In business, those mistakes and losses can add up to a lot of wisdom, enabling you to anticipate future setbacks or notice problems earlier, before they turn into major crises. As long as you apply the silver-lining strategy and use setbacks to come up with a better way, there's no limit to how far you can go.

At our USTA facilities, we've become world leaders in taking care of the needs of our players and fans. Each year, we use our IBM artificial intelligence (AI) technology (Watson) to collect all kinds of data to find ways to make the experience better, from how many staff members and volunteers we need on the grounds to make sure fans can navigate our facilities with ease to how best to launder the towels and how many to have on hand. What we don't want is sweaty players berating ball kids for running out of towels or not bringing them quickly enough, as Fernando Verdasco did at the 2018 Shenzhen Open in China when he was playing against Andy Murray. It's about anticipating what could go wrong so in those moments we can take excellent care of the players and fans at all levels. That's why we've become meticulous, leaving nothing to chance. We planned out every step an athlete or ball person might take. We thought through everything from the coolers for icing the towels during a match to the mat the ball kids put their knees on and how it will blend in with the court.

When I was a player, we didn't have our towels at the baseline. Since I wasn't always playing on a major court, we couldn't keep them hidden behind a clock or a chair umpire seat, where they wouldn't be a distraction to our opponents. It was incumbent upon us players to keep the court neat. Sometimes we could stack a couple on our chairs, but we certainly didn't have the benefit of kids running up to us with our towels while we were on the court—and I was a sweater. Instead, we relied on headbands, wristbands, or the ends of our shirts or skirts to dry ourselves off between points. But trends change. Today some players are almost ritualistic about dabbing themselves between points.

Sometimes our misses are a result of systems and rules that need to evolve. The USTA is an organizational behemoth, with processes and bylaws that require consensus and extensive due diligence to implement changes. It requires a lot of patience, which is not one of my strengths. But what I do have is a will of steel. I'm wise enough to figure out a way to turn a negative into a positive, recognizing that, even when I'm not able to get everything that I want, I can at least get things moving in the right direction. Fortunately, most of the people I work with respond well to my persistence. They've witnessed the way I deal with obstacles and realize I'm not going to walk away after one rejection or setback.

It's a matter of winning over everyone in the room to get forward movement on an issue. For the chairman of the board, ideas or topics often come up that require consensus. Making sure I had all the details and fully understood the end goal was paramount in speaking with or meeting the members of the board to answer any questions or clarify any information that would then educate the members fully. Although our board book materials are disseminated well in advance, the business is fluid, and challenges arise daily.

Being aware and knowledgeable is critical to getting the ball moving forward or even halting the progress of something that doesn't fit the mission. To bring others along, I made my case, teasing out different aspects of my argument from one meeting to the next. I also lobbied the holdouts individually. I would meet with them one-on-one to learn what their concerns were and assuage them. Understanding how cliques and alliances can form within organizations, I zeroed in on key members whom I knew had influence with the other skeptics, continuing to state my case one-on-one at every opportunity.

Underlying my persistence is faith. I inherited an unwavering belief in life's possibilities from my parents, who came out the other side of multiple adversities, including devastating health crises, with an optimism and strength that was stronger than ever. They taught me to keep pushing ahead, no matter how monumental my goal might be, and no matter the racism or sexism I might experience along the way. Seeing how they endured also helped me find my inner strength after the recent passing of my mother, a loss like no other. Through that experience I found an even deeper connection to Spirit.

We all need to have something untouchable we can believe in. My own strength resides in my Lord and Savior, Jesus Christ. People pray every day without even realizing it. Maybe they don't pray to God. Maybe they look for signs of support from a spouse, child, or sibling. But they're asking for strength and guidance all the same.

Ultimately, you can lean on your faith to get through loss. You can't put yourself in a winning position without connecting to something that's beyond the universe. After spending some time in South and East Asia, I began to meditate, learning how to sit in silence and focus on my breath to get in touch with that inner silence, that inner guidance. That silent strength and trust in something bigger than ourselves will help us get to where we want to go in life. Faith is a part of my culture, and I carry it like a beacon into every situation, professional or personal. Especially in today's unrelenting world.

6

BABY WOMAN

Dating a tennis player is risky—
love means nothing to us.

—ANNA KOURNIKOVA

When I was seventeen years old my father took me to Memphis, Tennessee, to play in the USTA Nationals, where I was running to the bathroom at almost every changeover. Like everything else that's troubling you physically and emotionally, when you're playing a match you don't bring it onto the court. You just play through the discomfort while focusing on the point you're playing. By then I was the seventh or eighth top junior player in the country and was trying to win a national title. But I was hemorrhaging. Add to this situation the fact that it was a steamy August day and my tennis skirt and shirt were soaked through with perspiration. Somehow, I was winning until the fourth round. But by then I was a physical wreck.

The next week I was in Philadelphia at the International Grass Court Championships. During a practice match I sprained my ankle. They put me in an ankle boot and the following day I was sent home. When I saw my mother, I mentioned my discomfort (something I felt awkward discussing with my dad). Alarmed that this had been going on for two weeks, Mom immediately took me to see my pediatrician, who had me admitted to the hospital that same day. An ovarian cyst had ruptured, causing havoc with my reproductive organs. After several hours of surgery, I woke up to learn that my fallopian tubes had been removed, along with one and a half ovaries.

In that moment I didn't comprehend the full meaning of what had happened to me. Going into the surgery, no one expected this outcome. It was only supposed to be a routine cyst removal. It was only after weeks of recovery that I began to understand the enormity of the reality that I would never be able to have kids, at least not through normal conception. I hadn't even had a proper boyfriend by that age. But seeing my parents' happy marriage, I'd dreamed of having a husband and family of my own someday. Inside, I felt broken.

If I hadn't been so driven to compete those two weeks on the road, I might have paid closer attention to the state of my well-being. I might have gotten a diagnosis earlier, before the cyst burst. It was an event that has hugely impacted the way I've conducted my life and relationships ever since.

Mine has been a life of sacrifice from the first time I picked up a racket. Not to be confused with loss, sacrifice is a choice. There is the path I chose and the one not taken. Whichever route you take, whatever kind of goal you set for yourself, there will always be something you have to forgo, because you simply can't experience everything while you are reaching for the top. Women especially are often forced to decide between career and a love life or family . . . but we've also proven that men aren't the only ones who can climb ladders, and we do it in high heels. That doesn't mean it's easy to maintain our balance. I decided to go in this direction knowing that there would be certain things other people get to experience that I would have to give up. It was not done to me, and I didn't necessarily regret making sacrifices for my career, at least not most of the time. As a professional athlete and career woman I had to make hard choices throughout my life in my quest to become the best. As a child I gave up hanging out with friends. As a teenager I missed parties and cookouts. As an adult, my travel schedule almost always required me to put career over long-term romantic relationships.

I was choosing tennis above all else because it was something I could not live without. I had a blast training, competing, and traveling. My tennis friends, who shared my goals to be the best we could be, became my world. When I got to compare notes with my childhood friends and classmates, I learned that my experiences were beyond anything they could have dreamed of for themselves. It's only through the lens of time and experience that I realized, "Ah, this is why I am single." I don't necessarily regret my decisions, but years later I understood that they could come at an emotional and, in my case, physical cost. Even though I made these choices in full awareness, there have been moments when I asked myself whether it was too high a price to pay. We must all make sacrifices to become successful. As leaders, especially women, we all face the prospect of having to give up something we love to stay on our chosen path. The trick is getting the balance right while feeling empowered and owning those decisions.

I didn't, always. Sacrifice was behavior I learned from my mother and father, who dedicated every penny to raising me and my two older brothers. The love we had for each other was profound, but we were never an especially touchy-feely, expressive family. We showed how much we cared through actions, and those actions almost always took the form of self-sacrifice. My parents gave up little luxuries to pay for my lessons and tournament costs. Summer vacations were sacrificed to go on the road with me.

Just how much they were sacrificing for me came to light years later, in 2006, as I was cleaning out my dad's filing cabinet. I found a handwritten ledger itemizing my expenses for the year 1983. They totaled over $35,000, which, in today's money would be roughly $88,000. I was in utter shock to discover exactly how much my tennis equipment, travel costs, coaching, and tournament fees had cost them. That tally was despite the fact that I had gotten a lot for free through the support of our community. Based on my parents' combined teaching salaries, which was $67,000, it's clear they made a huge financial sacrifice—but they did what they had to do to support my passion.

Of course, they never told me how much money was being eaten up by what at the time was my hobby. But I could sense their commitment in terms of energy and resources. My father, my true hero, worked three jobs early in his family life to make ends meet and ensure we never went without. He also cooked, did the laundry, and bought the groceries so that Mom could focus all her attention on us.

In return, I put away childish things. Early on, I made the decision to be the good daughter. Even before tennis was my thing, I observed the stress my rambunctious brothers put my parents through and vowed not to follow in their footsteps. Sacrifice was a learned behavior. I was determined to never be a burden to Mom and Dad. I'd be helpful at home, studious in school. I'd be a success so that they would never have to worry about me.

My mom and her friends used to call me "Baby Woman" because I always seemed like a miniature grown-up. They were joking, kind of. They had to constantly remind themselves that I was just a little girl. At my mother's funeral service my god-mother, Deloris "Dee" McJunkins, shared a story about the time I joined her and her daughter, Stacey, on a trip to Disney World. As soon as my mother said I could go I showed up with my suit-case packed. Dee couldn't believe how organized I was. I knew where all my belongings were. Every item in my luggage was in its proper place, the clothes neatly folded daily.

"We never had to worry about KK misplacing anything," Dee told everyone. "When it was time to go, she was always ready."

At a restaurant in Orlando, the maître d' handed us each a menu. I sat quietly staring at it when the waitress came over.

"Well Katrina, what would you like?" Dee asked me.

"I don't know, I'm only four," I politely reminded her. "I can't read yet."

The whole church erupted in laughter.

"She was so assertive and mature I completely forgot she wasn't even in school yet!" Dee told everyone.

It was as if I knew innately that I had to be the dependable one. As a toddler staying at home while my parents worked and my brothers went to school, I was the babysitter's little helper, direct-ing her in the kitchen so that she could find all the pots, pans, and other household items. When I was older, I became even more intentional about being the "good" girl. I was determined to excel in school and at tennis so that I could take care of my family some-day. I made that bargain in my head when I was no more than eight years old, and I've stuck to it.

Not that I felt in any way deprived at that age. I loved spending time practicing and playing tournaments. When I wasn't doing my homework or out on a tennis court, I had neighborhood friends to play jacks or jump rope double Dutch style with, or to beat at hopscotch. I joined the Brownies, but when my other friends were going into Girl Scouts, the choice for me was clear. Girl Scout meetings were on Saturdays at the same time as tennis practice. It was just another activity I had no problem letting go. I wasn't crazy about the green uniforms anyway.

At Whitney M. Young High School, a magnet school in Chicago where I missed overlapping with Michelle Obama by a summer, I had to forgo many of the parties and football and basketball games. But again, I was okay with it, because I was still doing what I loved. The school had a great tennis team where I won the Illinois High School State Championships my junior and senior years, being the first black player to do so. In addition, the school had some of the best academics not just in the city but also in the country, and I poured myself into both activities. When my high school buddies were heading to the mall, I was going to practice. Outside of tennis and my classes, I lived most of my high school experiences vicariously, through my friends. My friends filled me in after the fact on what I'd missed, so that at least I'd know how our sports teams were doing and have some clue about what was going on in our school when it was time for homecoming. (My senior year, I ran for homecoming queen and came in second. I wasn't thrilled, because I wasn't used to losing, but it allowed me to get closer to some classmates.)

But by my junior and senior years, I realized everyone was dating but me. Having skipped a grade, I started high school having just turned thirteen and graduated at sixteen, so in terms of my social and romantic development, I was already a year behind my classmates, which at that age is a big deal. Add to that the fact that so much of my spare time was devoted to tennis, my dating life was restricted to a few stolen kisses behind the bleachers, the gym locker room, or a secret rendezvous. Although there were a few guys I would hang out with occasionally, romantically, I never sought to have a "real" relationship. By the time junior prom rolled around, I didn't have anyone to go with. I went solo and asked Dan Duster, the class president, to take my official photo with me so that I wasn't alone. He obliged.

When senior prom rolled around, all my male friends assumed I must have a date. To avoid rejection, they didn't even bother to ask. By then I was making the news as a state champion and becoming something of a local celebrity. It must have made me seem unavailable. Two weeks before prom my seatmate in Division (homeroom), Barry Anderson, happened to ask me about my plans.

"I have no idea; I don't have a date," I told him.

"Are you serious? Will you go with me?" Barry asked.

I figured, "Why not?" We had fun that night, but nothing like in the movies, where the nervous date rolls up in a rented limo and pins a corsage on your dress. Barry was just a platonic buddy of mine, one of many. I was popular, vocal, and outgoing. But I was so involved in my sport that I felt remote from the rest of the high school experience, including my relationships. Sure, I got invited to parties on weekends. But knowing I would have to ask my parents' permission to go, then trouble them for a ride to the South Side, where most of the socializing was happening, it was just easier not to go, especially knowing I'd have to abide by my father's 10:30 curfew. It would have been too embarrassing to leave that early. Besides, I was usually too tired from practice or too mindful of the fact that I would have to get up early the next day to play.

My world expanded in college, where I formed fast yet lifelong friendships with Wendy Willis, Lori Shaw, and Lauren Lowery. I already knew Wendy from high school, but we grew much closer as freshman roommates at Northwestern. Through Wendy, I got to know Lauren, who'd been one of Wendy's besties since grammar school, and Lori, who was Lauren's sidekick at Robert Lindblom High School and my sophomore roommate. The cross-stitching of our sisterhood was already there.

The three girls had been doing a summer program on the Northwestern campus together a month prior to my joining them. That August and early September I'd been recovering from both the ankle sprain and having my insides all but scraped out. But my friends were already scoping out the campus, getting to know the jocks, the fun places to hang out, and dozens of other cool people from all over the country before the trimester officially started.

Campus life was everything I'd hoped it would be, with a ready-made social life. Finally, I was making up for being a step behind in high school. My girls were my support crew at all my collegiate matches, but since none of them were in athletics themselves, they were able to introduce me to the nonathletes on campus, broadening my base of friends to include those with a different set of experiences and perspectives. I couldn't go to all the college parties, because I still had practice and matches, but unlike in high school I wasn't just living around the edges of my undergraduate life.

I even met my first serious boyfriend, a Minnesotan who was there on a football scholarship. The sweetest guy, he could relate to my journey as an athlete. Just as dedicated as I was to my sport, he was all in with his, and therefore understood the concept of sacrifice. When you reach a certain level in a sport, you have that conversation with your coaches, your mentors, and yourself about what you're going to focus on. You might be great at multiple sports. You might love basketball, track, and baseball. But, again, you must choose. My boyfriend was the first person close to me who truly understood what it felt like to pour every ounce of your being into a single pursuit. He understood what it was like to have to decide what you're going to give up for the sake of an activity that demands your all. When we met, he was eighteen, only a year older than I was, but he was remarkably mature for his age and kept me grounded.

The next big sacrifice was the decision to leave school after my sophomore year. It had been only two years, but it was the most enriching experience of my life until that point, so tearing myself away from my teammates, friends, and college classes was one of the hardest decisions I'd ever made. Until that point, win, lose, or draw in my tennis matches, there was always tomorrow when I was an amateur in college. As long as I remained at Northwestern, I had a soft landing, with the unconditional support of family a short driving distance away and a team that always had my back. But this decision to go professional would thrust me into the real world of adulthood. Tennis would become my full-time job, and losing a match meant a loss of potential income. That was a heap of pressure to put on myself while still in my teens.

I didn't even want to think about the implications for my relationship. We kept it going long distance for another year before we finally accepted there was no long-term future for us. He had a job waiting for him back in Minneapolis after he graduated. It was time to have the conversation.

"Kat, I know what my path is, and I know what your path is. I don't want our relationship to turn into a hindrance to your career."

This was my first true love. My parents were also college sweethearts. They got married and built a life together and I'd always wanted what they had. But I wanted this life in tennis more, and he knew it. Today he's one of my dearest friends.

When I was twenty-five, after a few years on the professional tour and one or more disposable boyfriends, I fell hard for a man in my home base of Houston, where I had moved after my first year on the professional tour. The connection was mutual, instant, and intense. He was a wealthy investment banker, but he'd also been a college athlete who'd aspired to go professional until he was forced to make a choice. Again, he understood the competitive environment I was in and the level of commitment I had to make. He admired what I did and was supportive of my dream, to the point where I felt I could share everything with him. I mean, everything, including the story of my surgery, which less than a handful of people knew about. It wasn't like me to be so open with anyone. But this was one of those rare connections between two people where you feel like they really see and accept you for who you are.

We spent every spare minute we could together. During this hot and heavy time, I couldn't lose. I was on fire on the court. Off court, I was giddy. When I was on the road, I couldn't wait to get back to my hotel room, where we'd talk for hours on the phone, often into the wee hours of the morning. It limited my hours of proper sleep to perform the next day, but I was winning!

My man also had to travel for work, so he understood the crazy lifestyle I was living. But it wasn't enough. He wanted what he wanted, and he hated coming back home to Houston to find I wasn't there. Instead of picturing what might be possible for both of us down the road, he focused on what he couldn't have in his immediate future.

"Kat, I would never ask you to cut your career short just for me," he told me. "But I need someone to be there for me in the here and now."

"We *can* have a life together," I assured him. "Even though I have to travel, my professional career isn't forever, and I'll always come home to you in Houston."

That wasn't the whole story. He wanted a family. He had been married before and divorced without children. It would be a few years before I could even attempt the process, although I was more than willing. He was not, because he'd witnessed the disappointments that friends of his were going through with IVF.

"But what if it takes the first time?" I asked him. "I'll still be young when we start trying. We both will be. It can happen for us."

He didn't want to wait to take a chance. The timing was wrong for both of us, but more so for him, and it was his choice, based on his own needs, to end it. I was devastated. This was the first serious conversation about my fertility I'd had with someone I'd envisioned spending my life with.

Although I know it wasn't his intention, his rejection made me feel less than. Months later, I fell back into another long-term relationship. But there was always something stopping us from fully committing. This next boyfriend always feared I would break things off, when in fact, I was looking forward to getting engaged. While I was playing in Europe, we communicated daily for this particular period of three weeks straight, then crickets. For the next week I was confused. Why the silence? Did something happen to him? I knew his mom had been ill; maybe she had taken a turn for the worse? I got home to a breakup letter in my mailbox (no doubt if this had been in the days of cell phones, it would have been a text). But I have no regrets because that breakup allowed me to continue on the path that was meant for me.

Those two romantic disappointments made me move forward into subsequent relationships with my eyes wide open. To guard my heart, I'd end things at the first sign of a problem. I wouldn't even give them the opportunity to correct course or explain themselves. I'd move in for the preemptive strike in the same way I served and volleyed at the net to shut things down. It wasn't always the right thing to do, but it was easier for me. I could be on the

road to Europe, Australia, or Asia the next day, distracting myself
with tennis. Whatever issues I had could be worked out on the
court, which was my office and therapy space, or so I thought.

The sacrifice at this point in my career was a rich emotional life.
Shaking it off and moving forward from past disappointments had
become one of my specialties. But it wasn't always healthy, because
I wasn't talking through and processing everything the way I
needed to. It got to the point where there was this hard kernel of
truth buried so deep within that even I couldn't access it.

I used to confound my coach, Willis Thomas. When I was up
against the top-seeded players in singles matches, he could see I
was more than capable of beating them. Physically and in terms of
my skill set, I was often their equal. But something was missing.
I'd play really well against the top players, then choke in the final
set. As my coach, Willis's job was to prod until he could find out
what was holding me back emotionally.

When you are a professional player, you must be as free of everything as you can be when you are on the court, because if anything
negative creeps in your mind it can immediately impact your pro-
ductivity as a player. It may not be obvious to the casual observer,
but professional players and coaches recognize that something's
going on when they see one too many unforced errors.

That's what Willis saw, but I wouldn't give him the satisfaction
of letting him in. Exposing my vulnerabilities to my coach would
have required me to admit them to myself. Even when it was obvious something was terribly wrong, like in 1992, when I was playing in Wimbledon with another ovarian cyst, which had turned
into a tumor the size of a honeydew melon. When I lost in the
second round of singles, I refused to share what was going on with
me. I kept on playing through to the fourth round of doubles, but
I was in agony.

I was always getting cysts, but this one was a freak show. It showed up at the peak Grand Slam season, so I didn't want to deal with it and risk missing out on some of the most lucrative tournaments of the summer. It was another sacrifice to my sport, I suppose. I did my best to support and hide the monstrosity with a compression bandage. Fortunately, in those days the style was oversize clothes and not form fitting like today. I also tried treating it with acupuncture, although the pain persisted. It was all I could do to play through it and not show my discomfort to my opponents. My mindset was to keep going until it was time to return to Houston, where I would have the long-overdue surgery.

The tournament doctor and trainers knew about my condition only because they had to, in case of a medical emergency. It pretty much was at that point, but when you are just trying to endure, denial helps. As soon as I lost my last match that Monday and could go no farther in the tournament, I called the doctor to book the surgery for that Friday. I was in the hospital for the next two weeks recovering. Admitting how serious my illness was during the summer circuit would have invited pressure to skip one of my favorite Grand Slams. So, when Willis poked too much, he'd get the same result: I'd snap at him, then walk away.

Whenever he suspected something was going on with me, Willis would try a work-around, gleaning pieces of the story from my girlfriends who, when time permitted, came to visit me on the tour. Wendy, Lauren, and Lori were the only people who knew me well enough to give him some insight. Although Zina and I were like sisters, he couldn't ask her because, close as we were as doubles partners and friends, there was always that slight competitive edge between the two of us. Willis couldn't ask Mom either, because he was a little afraid of her.

"Your mother had such a regal carriage," he shared with me,

years later. "Something about her demeanor told me to mind my business."

Not sharing with my mother is my biggest regret. I know that she always wanted me to come to her about anything, but I had convinced myself that the more loving, less selfish thing to do was protect my parents from anything that might make them feel any added pressure, disappointment, or concern. We went through our lives protecting each other from information we thought might be hurtful, but that cost us, because we deprived each other of the opportunity to draw even closer.

My parents did the same thing to me. When one of my former coaches, Chris Scott, died of a heart attack at just sixty-two, on the Hyde Park Tennis Club court in October 1989, they made the decision not to tell me. I was playing a tournament in Europe and they feared I'd drop everything to come home and pay my respects. I discovered everything when I got home, but the memorial service had happened days before. For months afterward, I was furious with my parents for not telling me. Whether I cut short my playing to come home or not should have been my decision.

It was neither the first nor the last time they took pains to "protect" me from facts I would like to have known. But I understood their intention. Misguided as it was, the withholding of information was done out of pure, selfless love. They were sacrificing for me, as I was sacrificing for them. The problem was our sacrifices canceled out my ability to freely choose.

If my parents had been more open with me, perhaps I'd have been more transparent with myself. But I am a product of how I was raised. It's only as I've matured as a leader that I've realized the importance of managing my emotions versus stuffing them down. As an African American woman, protecting my privacy has always been paramount. It's because I don't want the perception of being overly assertive to distract from the matter at hand. I walk a fine line between self-consciousness and deference to the feelings of others. Self-awareness is everything as you seek to find that balance, which is something all aspiring leaders need to cultivate if they hope to build a culture of transparency.

To that end, in 2014, prior to taking on my leadership role at the USTA, my business mentor Jim Kelly recommended that I meet with Dr. Lily Kelly-Radford, a respected executive coach who also happens to be his sister. Jim knew me well, and he understood the responsibility that I was walking into with the USTA. But, as someone who has always been so self-reliant and reluctant to open up, even with my closest friends and family members, this was a huge step for me. After doing some research and learning how many top leaders in business rely on executive coaching, I decided to set up some face-to-face sessions with Dr. Radford. I figured speaking to an objective third party with emotional intelligence, or EQ, could help me shine a light on my blind spots and give me permission to be more vulnerable. Turns out, she was extremely helpful and continues to guide me today.

As leader of the USTA, I realized I no longer had the luxury of hiding behind my armor. I had to become more accessible to the members of my team who needed to trust that they could let their guard down, because God forbid anyone should feel like they have to censor themselves. I never wanted to find myself in a situation in which, however well meaning, people were keeping inconvenient facts from me. They needed to be able to bring anything to my attention without fear of judgment or reprisals so that concerns could be dealt with before they manifested into bigger problems. In my interview, it was brought to my attention that I was perceived as unapproachable. I told them it was nonsense, and that if others felt they couldn't approach me, it's their problem, not mine. I recognize that my look may not have always been welcoming, a look I acquired from my competitive days, but I was very approachable. I have since tried to relax my look a bit. But anyone who ever ventured to engage with me was greeted with open arms. If you don't ask, you don't receive.

I allowed myself to be particularly vulnerable with Andy Andrews, whom I'd nominated for the role of first vice president when I was ascending to the USTA leadership role. Andy, a former professional tennis player from Raleigh, North Carolina, owned a real estate development company, so he seemed like the perfect person to have in my corner during both the USTA Billie Jean King National Tennis Center upgrades and the development of the USTA National Campus in Orlando from nothing but cow fields. These were to be two of our organization's biggest milestones in decades, both being handled under my watch. Our leadership team had done all of the groundwork in partnering with the developer Tavistock in Lake Nona to make this happen. There was to be a hundred courts of all kinds—full-size and kid-size—with every surface except grass, making it the largest, most state-of-the-art facility of its kind in the world. Yet I knew nothing about construction, and I was candid with Andy about my lack of expertise.

"Andy, I'm counting on you to give me a crash course in construction," I told him. "The lingo is completely foreign to me, so I'm going to ask you a lot of dumb questions."

Exposing myself like that with someone who was next in line for my position may not have seemed politic. But it was the best thing I could have done, because Andy took it upon himself to patiently tutor me, feeding me questions to ask the developers, breaking it down for me when I had to bring other board members up to speed; never BSing me when there was a snafu; and remaining calm, steady, and reassuring throughout the process. He would even fly up to New York from Raleigh for the day on his own dime, giving his personal time and resources just so that he could be by my side in meetings with developers or represent the USTA if I was traveling. Entrusting him in this way helped us to keep these projects on track and on budget.

"Andy, your steady, dependable presence has allowed me to not gray as fast as I otherwise might have," I told him. "From the bottom of my heart to the top of my head, thank you!"

We all make sacrifices, whether they relate to our careers, our education, our family, or our faith. There isn't a soul on this earth who doesn't look back with the wisdom of experience and wish they could get back something they gave up in order to make it all the way to the finish line. That's okay. We're human. It's what we do with that information going forward that matters.

My biggest sacrificial lamb turned out to be my closest personal relationships. Long after my professional playing career has ended, I am *still* single. Each new challenge I take on requires more of my time. Even after my tenure as head of the USTA I am on the road more than half the year. If anything, the amount of time I spend overseas has increased as I get more involved as vice president of the International Federation of Tennis, while maintaining my NJTL program in Harlem.

My career is what it is, and I'm not slowing down anytime soon. My lifestyle over the past couple of decades has not exactly allowed me to put in the necessary time to maintain a strong romantic relationship. But over the years I've also discovered it's relationships we have with friends and family that matter most. Life is challenging enough without being part of a dysfunctional couple. Forcing it never works. I've often been teased that I never married because I was always looking for a man like my father: a gentleman, a provider, and a kind soul. I'm not so sure those qualities exist anymore. But whether they do or not, I know I am going to be just fine. But when that person comes along, I'll be ready.

My father passed away on May 8, 2020, a day before what would have been my mother's eighty-fourth birthday. He died of a broken heart. It was a major loss to my soul, losing my true hero just nine months after my mother's passing. He was extremely proud of me, and I owe a great debt of gratitude to him for the many sacrifices that he and my mom made for me. Maybe a love interest will come my way and remind me of the inner strength that he had.

7

NO WOMAN IS AN ISLAND

*However great your dedication, you never
win anything on your own.*

—RAFAEL NADAL

t feels as if I've been receiving the gifts that tennis has given me
with open arms from the first moment I picked up a racket.
Making a difference and succeeding in anything, takes courage,
stamina, perseverance, self-confidence, and discipline. All skill sets
that are learned through the sport of tennis. Tennis changed my
life in the doors it has opened for me. My experiences on the court
provided lessons beyond the classroom, such as being prepared to
own any room I might walk into.

One autumn morning in 2016, I was sitting across the table from
Billie Jean King enjoying a hearty brunch of waffles at Barawine,
near the 369th Regiment Armory HJTEP facility. Billie Jean, one
of the great sheroes of our time, has done so much for women's
rights, demonstrating by her own actions how much we can achieve
by working small scale and one-on-one, or leading monumental

movements. As an advocate for gender equality who has fought hard for social justice, I have admired her since I was a young girl. So I paid attention when she fixed me with her steady gaze and asked, "Well, what's next?"

It was the beginning of my second two-year tenure as head of the USTA, the first time in history anyone had been given another two years in the job, and I'd been more preoccupied with the question: "What's now?" I was still catching my breath and thinking about everything I needed to get done while I was still in my role. Count on BJK to remind me that I should continually be setting goals for myself, aiming high and believing I could exceed them.

BJK has always been my sage in sneakers—a caring voice who never once told me what to do. Instead, like the best coaches in any sport, she always encouraged me to think for myself. I looked up to her, but just as important, she never looked down on me, always making me feel like I belonged. She's been a constant guiding light.

I've known Billie since before I turned pro. She was living in Chicago with her partner, former doubles champion Ilana Kloss, and the couple used to play tennis at Midtown Tennis Club, where I often practiced. We were all hanging out one day after a workout when I got up the nerve to ask her to talk to my communications class at Northwestern. Not only did she show up to inspire a bunch of freshmen, she's turned up at pivotal moments in my life ever since.

I've been fortunate enough to come across many other mentors along the way, from my mother and father to the coaches who volunteered their time to instruct me both on and off the court, sharing values not just to make me a better player but a better person. In the beginning, we didn't have the money to pay for their time. My first coaches found me and chose to share their wisdom and experience. If they hadn't, my potential for tennis could have died on the vine, because raw talent only gets you so far.

Recognizing that I was being given a gift, my mother and father made it clear that I was to pay close attention. Ever since then I've looked for opportunities to learn from other people and situations. It's a mindset that's served me well throughout my career. We tend to think of tennis in terms of hitting the ball hard to our opponent and letting them have it with an aggressive style of play. But you must also know how to tactically receive the ball, to respond appropriately.

It's believed that the etymological origin of the word *tennis* is the Old French *tenez*, which means "to take or receive." Winners of the game do not just rely on a power serve. The consistent champions also know how to receive the ball. Andre Agassi was one of the best returners in the game—seeing the ball clearly off his opponent's racket with great anticipation to move left or right. One of the greatest strikers of the tennis ball. They have anticipated what is coming and they are ready for it. It takes years of practice, dedication, and awareness to know precisely where it's coming from and how best to return it and position yourself to win the point. The same holds true for wisdom. I've been blessed to have dozens of great coaches and mentors over the course of my tennis and business career. They've come in many guises, and it's often felt as if they found me. But when they showed up, I soaked up every drop of knowledge they had to share with me.

My first coach was Tony Fox, who, along with Kim Williams, was one of the instructors at the youth program in Garfield Park. Tony became so close with Mom and Dad that he became like a third parent. My folks made it clear to me that, in their absence, Tony's word was law. Under no circumstances was I to disobey.

"Katrina's gonna be with you all day, so we want you to know she has to mind you the same way she minds us," Dad told him. "And you have our permission to discipline her if you need to. We are telling you this with her listening, so if you get any back talk or sass, let us know."

One Sunday afternoon I was at the Boys Club gymnasium doing laps and crying. Tears were streaming down my face and snot was running down my nose. I'm not sure why I was being made to run. Either I was being punished for something or it was part of my physical conditioning. Either way, I hated it. Mom was sitting on one of the benches grading papers when a lady came up to her.

"Is that your daughter?" she asked her.

"Yes."

"Don't you know she's crying?"

"Well, she's with her coach, and if he's not worried, I'm not worried."

Finally, Tony came up to me.

"How many laps is that, Katrina?"

"Six!"

Tony laughed. "More like four. You've got six to go. In all those tears you got confused."

Immediately I stopped crying.

Although I never talked back, I wasn't a perfect angel. Occasionally, I'd try to sneak something past him. During the summer of 1976, Tony ran an inner-city tennis youth program, Youth Action, for kids from all over the area, all minorities. The point was to teach us not just tennis but also life skills. When we got out of line, Tony had fenced in a little area by the clubhouse window that we called "the jail." These were the days before time-outs, but it was a similar concept. Tony would have us stand there quietly for twenty minutes. It was torture, because we were in plain view of the courts. Nothing was worse than having to watch from afar as everyone else enjoying themselves playing tennis.

I was goofing off, neglecting to do a drill, when I got sent to jail. It was five minutes to 4:00 p.m. when my father was due to pick me up, so I figured I'd gotten off lightly. All I had to do was let the clock run out. But Coach Tony must have caught the gleam in my eye as I watched the second hand tick by. When Dad pulled up, Tony went out to meet him.

"Bull," which is Dad's nickname, "you need to park and come inside. Katrina has another fifteen minutes on her jail time."

It wasn't all about discipline. When we'd done particularly well in our lessons, Tony used to take us all out to eat at different restaurants in the neighborhood. Wherever we ate, we were never allowed to order junk food like hamburgers or fries. We had to read the menu properly and order at least one vegetable. The boys had to pull out the chairs for the girls, and I got told more than once to keep my elbows off the table. It was all designed to instruct us in the ways of nutrition, manners, and dignified conduct when we went out into the world. To Tony's credit, every child in that program ended up in college; a couple even earned PhDs. Youth Action was a pilot National Junior Tennis and Learning (NJTL) program, similar to the Harlem Junior Tennis and Education Program, and the network has grown to almost three hundred chapters nationwide, producing some incredible players and people.

Tony was a born teacher. Like me, he wanted to play tennis when he was around five years old, but as a boy coming up in 1950s Chicago, "tennis was a girl's game," so he wasn't allowed to play. A far cry from today, where tennis is an equal-opportunity sport, the people around him didn't feel he belonged in that room, so he took up swimming instead. As a young man he started taking tennis lessons from a prominent pro, who taught him for free on the condition that he would teach what he'd learned to someone else. Tony took that to heart, teaching as he learned. Soon he discovered that he loved teaching more than he loved playing.

Tony didn't generally believe in starting someone as young as I was in athletics, but he believed in my potential and made an exception, giving up his Saturdays to train me on the gym floor at Washington Park Field House on the South Side. That's where I learned to handle the faster balls, although Tony was a stickler for turning me into a well-rounded player who could adapt to any circumstances.

He was also a proponent of "internal learning." In other words, instead of telling me what to do, Tony would put me in situations that would require me to come up with the solution or apply the technique, creating lessons that I would never forget. Tony jokes that in today's world he'd never survive, because his teaching methods would be viewed as borderline abusive. Maybe so, but they sure worked!

When he was teaching me how to protect my body with a backhand volley, he fired shot after shot at me. I was getting pummeled.

"Okay, the choice is yours," he told me. "You can either position yourself with the racket in front of you, holding it tight with your backhand volley, or you can keep getting hit."

Tony showed me how to turn a negative situation into fuel for motivation. I quickly figured out how *not* to let the ball thwack me. Getting hit helped me to internalize the lesson. Years later, Tony admitted that the balls he was hitting to me "had a little pepper on them." But today I am grateful for the pain because these drills enabled me to become a great volleyer.

Again, great mentors fire you up so that you can achieve more than you thought you could. They remind you of who you are and what you are capable of, accepting no excuses for falling short.

When the arena is full of distractions and everything seems to be going against you, they whisper something in your ear that enables you to focus on your goal. This is the approach I take with those who I mentor. I often receive an email from someone, mostly young ladies, who would like for me to mentor them. My goal is to listen, understand their goals, and ask pertinent questions that allow them to figure out the answers on their own. Being an ear can be the biggest asset to others. This allows me to really hone in on what is important and realize how I can truly assist someone while building their confidence.

Navigating the waters as a black woman can be tricky. It can be an asset for some looking for diversity in finding a woman of color—a two for one. Or it can be challenging in trying to ascend the ladder where people may not be so keen to answer to a black woman.

I was eight years old when I noticed that minority children were either slated to play against each other or put up against the top seeds in the early rounds. I don't think it was a coincidence. It was my opinion that these suburban tennis clubs didn't want to see a player of color beat one of their kids. They engineered it so that we would be knocked out of the tournaments early. It upset me.

"The only way for you to eliminate that is to become a top-seeded player," Tony told me when I asked about minority players being pitted against each other in the early rounds.

So, I did. With my determination to show them, I got steadily better, until I became the No. 1 seed in the Chicago District Tennis Association Junior events. I was the only black girl in the tournaments, which was significant. Tony continued to encourage my tenacity and taught me the mechanics of my serve—my ultimate weapon. A couple years later I was with another coach, but Tony came to a tournament at the Midtown Tennis Club to watch me play. He was standing in the lobby when he heard a bloodcurdling scream. It came from the direction of a girl standing in front of the draw in tears.

"What's the matter honey?" her mother asked.

"Adams!"

"Well, dear, just do the best you can. It'll be a good experience for you."

"I'm gonna get killed!"

Tony was still chuckling when he told me after the match.

"That was some perfect circle moment!" he said.

Tony demanded accountability. Win or lose, he made me explain what happened after each match. It was never just about the score. It was about what I had done to win or what my opponent had done to lose. True winning was based on my playing well, not my opponent playing badly, because, as he questioned, "What do you do when your opponent changes tactics and takes the lead?"

Tony retired a few years ago to take up photography full-time because he says that "kids these days are a little too spoiled and narcissistic for my taste." That may be a little harsh, but the culture of training has changed from his old-school approach, and many parents are quick to intervene at the first sign of their child's discomfort. Tony also believes many kids are overpraised.

"You don't get a ribbon for washing your face and brushing your teeth," he likes to say.

I responded well to his brand of tough love. When he changed clubs, Mom asked me why I wanted to follow him all the way across town.

"Because, when Tony tells you you've done something right, you know you've done something right," I explained. "You won't hear those words until you do."

Like all great coaches, Tony recognized when I had surpassed all he had to teach me. He took the step of bringing me to his mentor, Chicago sports legend Chris Scott, who was known for his work promoting tennis on Chicago's South Side and turning players into national champions. For years he was also the men's and women's varsity coach at the University of Chicago.

A World War II veteran and Harlem Globetrotter, Chris was a different kind of coach. He took up tennis later in life, learning his skill by watching old tapes of matches and reading instructional books. The rest was common sense. That machine-gun technique that Tony used to get me to position my racket and body better came straight out of Chris's playbook. Chris was the coach who helped me develop my mechanical base until I turned pro. But one of the biggest lessons I learned from him was endurance.

He used to wheel an old shopping cart full of balls onto the courts. There had to have been at least three hundred balls in that thing. He started by shooting fifty of them side to side across the net. Just when I thought I had no more gas left, he'd fire another ball, then another. We finally got to the point where he was feeding me all the balls in the basket. I felt the burn in my legs and my gut. Every muscle ached. I was crying and complaining the whole time, but not once did he raise his voice or show frustration with me as I was slowing down. Chris kept firing those tennis balls. It got to the point where I felt numb, with limbs as heavy as lead, and I was forced to dig deep for that last burst of energy. Somehow, he knew I could go for another twenty minutes before I did. He helped me to discover a level of tenacity and endurance I didn't know existed. When I came out on the other side of these ball-feeding marathons, I was stronger, better, and faster.

Another early coach, Rod Schroeder, who helped me get to the next level in my playing, ran the Rod Schroeder Tennis Academy at the Tam Tennis Club in Niles, Illinois. When I was ten, my parents heard about Rod, who had a reputation for developing junior players at the national level, so four days a week they drove me the twenty miles northeast to receive lessons at the suburban club with a group of players who were my age and older. Other days, I continued to practice at Hyde Park Tennis Club or Midtown. Rod noticed my serve-and-volley style as well as my ground strokes. Having seen the way I made high school–level boys and girls run all over the court, he decided it was time for private lessons. Before long, he became the coach who traveled with me all over the country for championship tournaments. Rod and his wife, Myrna, grew so close with Mom and Dad that when it came time for me to make my decision to turn pro, he was the one I asked to help me break the news to them.

When I turned pro, I found a coach whose approach was as in your face verbally as one of Chris and Tony's practice cannon balls. Willis Thomas, who was black, was Zina Garrison's coach on the professional tour. When Zina and I became doubles partners we shared him. But it was a love-hate relationship to begin with. If I lost a match, or failed to play as well as I should have, Willis would say, "What the hell was that?" As if I didn't feel bad enough. I'd respond with, "#%@* you, Willis!" and walk away. I never gave him the satisfaction of agreeing with him or immediately acting on his advice. But over time I came to really appreciate his blunt style. We could be aggressively honest with each other knowing neither one of us was going to hold a grudge. We all need someone in our lives who's never afraid to tell us the truth.

Looking back, I realize my parents had great taste in mentors. Perhaps because they themselves were teachers, they recognized people with shared values and a level of experience and insight from which I could benefit. They created a village. (My personal board is my village today.)

One of those earlier villagers was my second-grade teacher at St. Mel Holy Ghost grammar school, Emma Teagues. Because my parents were teachers at schools that finished classes later than mine, Mrs. Teagues agreed to keep me after school until my dad could come to pick me up. Well-read, refined, and well-traveled, Mrs. Teagues must have been in her sixties when we spent our time together, quietly reading the books that lined her shelves or talking about her trips around the world.

My biggest takeaway from Mrs. Teagues was the value of reading. Even if I didn't have homework to do, she would hand me a book, so that in the thirty or forty minutes waiting for Dad I was getting ahead in my studies. She also taught me how to be an assistant, tidying up the classroom, cleaning the chalk pads, or organizing some papers for grading. Being an assistant teaches you how to be a leader, because it helps you recognize how to guide and communicate with someone.

My teachers weren't always older adults. Sometimes they were fellow players much closer to me in age. They were teenagers and young adults who were in it with me, practicing and sometimes competing directly as we played matches, daring me to be better.

One of those young early mentors was Mel Phillips. Mel was one of the stronger junior players in Tony's Youth Action program. He was seventeen and I was ten when we started playing together. Years later, he shared with me his first impression.

"They were feeding balls to you and you were running side to side, hitting ground strokes, and it was just so effortless. You freaked me out! Who was that kid?!"

Later, when I was about eleven, Mel was playing in a Mixed Doubles tournament at his hometown court in Maywood when they brought me in as a young ringer. He'd been dominating until I showed up. It was the first time in a doubles match that he started hitting the ball toward the guy on the opposing team. In Mixed Doubles, picking on the female player was the standard strategy, especially in park tennis.

Mel was sufficiently humbled, later only too willing to become one of my practice partners. That was a gift, because finding a strong male player willing to play a much younger girl back in the day wasn't easy. He was self-assured enough to handle getting beat by a girl, and that generosity of spirit helped me develop. Today, Mel is like a member of the family. He went as far as he could in the amateur divisions, and now he works as a life coach and a school social worker in addition to running a youth program in Maywood.

Tyrone Mason is another tennis buddy who, like Mel, took an interest in my development. He picked me up and dropped me off whenever I needed, took me to practice and wiped me off the court. I was persistent and wanted to play often, and because I was getting better, I began to win more games. That's why, as players, we come back to the courts every day. We want to see how far we can push ourselves, how much further we can go, and how much we can learn about ourselves. We want to measure our progress with both wins and behavior management. I never looked back when, at sixteen, I finally won a match against him. Of course, Tyrone never made it easy. Instead, he showed me how to fall and get back up for the next battle. He didn't care if I cried, screamed, or yelled after losing points. He just kept teaching me lessons of self-reliance. As we continued to play, my goal was to beat him worse than the previous match. Tyrone was like a brother to me, so those battles were tough love in action.

Donald Young Sr. was the player who truly gave me fits and taught me how to be patient and tactically change my strategy. Donald was the biggest pusher. He ran everything down and ran me all over the court. It took me years to finally beat him, but each match I won one game more than the previous match. You learn from your losses and Donald taught me tons of lessons. I had to be more fit, patient, resilient, and aggressive, skills that I needed when I went to the next level.

I didn't get to where I am alone. Those mentoring moments didn't necessarily go on for years. Because I was young and developing, even the smallest deed could have a profound impact on me in that moment or situation, propelling me toward something greater. I was fortunate to find these angels all the way along my journey.

Experience was another great teacher. When I first joined the WTA Tour, traveling the country and the globe, there were many times I had to figure things out for myself: how to pack, how to book planes and hotels, how to manage my expenses, making sure my documents were in order, never allowing my passport out of my sight. Being one of four or five black players, I was always looking for my comfort zone with my coach, Zina, or Lori.

I'd learned to be somewhat independent in college. But back then, I was close to home and surrounded by my close friends, including my roommate and best friend from high school, Wendy, and my college mates, Lori and Lauren. Of course, there are other players on the tour, some more friendly than others, but they were also the competition. Being thrown together with women as if we were living in a sorority, enjoying the camaraderie while never forgetting they were my enemies on the court, was like a crash course in managing complex work relationships.

However, many of my closest friendships on tour were with foreigners, mostly from the Netherlands. They never saw me as different because I was black. They appreciated my ethnicity and simply treated me as a human being. This gave me great pleasure and freedom to be me, unconscious of race. It's interesting, although I never looked for racism or felt I was being treated otherwise, I always felt that I had to be somewhat different with my white American friends than with the Europeans. I really felt comfortable being myself with these players.

Race in America was always a divider. Tennis was a predominantly white sport where I had to fit in. I was always viewed as a "special talent" because I was black. If I were white, I would have been one of the "girls." But competition was different for me. I always felt like I was there with the same goals of winning like everyone else. Plus, in the United States, my peers and I were fighting for bragging rights as the highest ranked player, which had nothing to do with race. That's another reason why it was easier to be friends with the foreigners.

Doing what you love and seeing the world sounds glamorous, and there were many times when being in a different country or city every weekend was a blast. My first time playing in Melbourne, for example, we all went out to celebrate a fellow player's birthday. We partied at a nightclub called Metro until 6:00 the next morning, just in time for him to make his 9:00 a.m. flight back to the US, where the time difference would allow him to celebrate all over again! As one who usually turns into a pumpkin by midnight, I'd never done anything like that before. I lost a day's practice before the next tournament in Tokyo recovering from the revelry and one too many beers. But it was my first year abroad, fresh out of college, so I was entitled to a few wild moments.

Later that year, I played in Moscow when Russia was still the USSR. I walked around Red Square, trading dollars for rubles on the black market, buying fur hats and nesting Russian dolls. Less fun were the accommodations: a hotel where the tap water came out brown. There was little we could drink or eat that week. All the food was swimming in butter, from the eggs to the chicken Kiev that was one of the few edible things on the menu, washed down with bottles of vodka—the only liquid they served with the meals. The water available to us in the tennis stadium was carbonated, so if we dared to drink the stuff, we would inflate into Butterball turkeys and cramp up in the middle of matches. My mom traveled with my because she was afraid to let me go to Russia alone. She was a nervous wreck when I told her I was being a tourist. That Friday the US Embassy invited us over, and we descended on their commissary like a pack of hyenas, scooping up bottles of Evian water, American cereals, yogurts, and peanut butter to take back to our hotel rooms. The whole time, it felt like someone was watching us, which they were! It made me appreciate all the freedom, comforts, and access I had back home. At least my mother was with me.

These were life experiences I would not trade for anything. But it wasn't like today, when there's more money for players on the tour; unless you were one of the top-ranked players, only a few had sponsors to assist with travel expenses, coaches, or personal assistants. We didn't arrive somewhere with entourages tending to our every need. You earned your keep by winning. The more you won, the more you made. The middle-of-the-pack players were just trying to break even. If we had a great week, we made a profit.

That's still the case for lower-ranked players, although conditions have improved. In my first few years, we had to pay for our own hotels. Over time, tournament organizers started offering hotel accommodations, as long as we were winning. They later gave us a couple days' grace period after we lost, so that we could practice and train for the next event. We cut corners wherever we could to hang on to our winnings, often sharing rooms and putting our coaches in the other paid room to save on out-of-pocket expenses. We even shared coaches.

We had to fight for practice court time, often at crazy hours, because the stars on the circuit got first dibs. If you didn't have some sort of sponsorship deal, you didn't stay in five-star hotels, unless the tournament was being especially generous. We were responsible for making sure our passports were up-to-date or ensuring we had the right visas. Planning and preparing for the next trip, and the next, became like a second job. As players we were constantly seeking ways to make our itinerant lives less stressful. Word got out about a travel agent who could book us around-the-world tickets, enabling us to fly first class domestic and business class overseas at a reduced cost. These tickets also allowed us to make last-minute changes at no additional costs, which was a godsend, since our flight schedules changed at the last minute depending on whether we won or lost a match. We later found out that what he was doing was illegal, and he got arrested. We were lucky to have had access to those kinds of travel deals while they lasted.

Mel, who by this time was working in the airline industry, used to drop in to watch me play at some of these events, and he was taken aback by what a grind it could be. You were basically thrown out there, landing in strange countries where you don't know the language and customs. I had to be hyperaware, paying close attention to nonverbal communication in order to feel my way through.

Those situations became my classroom, helping to finish the education I'd cut short at Northwestern. Peachy Kellmeyer was one of the "professors" who helped me to navigate this new reality.

Peachy has been a force for women's tennis since she won the Orange Bowl championship in 1957. At fifteen, she became the youngest player invited to play at the US National Championships at Forest Hills—the precursor to the US Open. In college, at the University of Miami, she became the first woman in history to play on the men's Division I team. But her achievements since retiring as a player are what truly paved the way. As director of phys ed at Marymount College in Boca Raton, Peachy instigated a lawsuit that would dismantle a prehistoric rule prohibiting athletic scholarships for women. The ultimate result was Title IX in 1972, which stated that, "No person in the United States shall, on the basis of gender, be excluded from participation in, be denied the benefits of, or be subjected to discrimination under any educational program or activity receiving federal financial assistance." I am a recipient of the seeds they planted.

In 1973, she became the first employee of the WTA as a referee and was soon tapped to be tour director of the WTA Tour. She spent the following decades building the WTA into a global platform for women players, bringing our Virginia Slims year-end championship event to Madison Square Garden in 1977. Year after year, Peachy pushed for increases to prize money, from a finals prize money purse of $100,000 to $7 million. She also spearheaded the WTA campaign to achieve equal prize money for women at all four majors, which finally happened in 2007. Peachy's career contribution saw her get inducted into the International Tennis Hall of Fame as a contributor to our great sport in 2011, alongside Andre Agassi. As Virginia Wade once said, "Peachy is the WTA."

It wasn't enough that we got paid fairly and were given a decent platform for our sport. She cared and wanted to make sure we were happy on the tour. If there was a problem, I knew I could confide in her. Peachy wanted nothing more than to see me flourish. She gave me media training and help with the fine art of diplomacy. We could be talking to ball kids one minute and the president of the local bank the next, so you had to learn to adapt and engage with people from all walks of life.

At various locations we were often invited to cocktail parties with local dignitaries, and Peachy would give me primers on who some of the guests were, as well as tips on local customs and safe talking points. I cared deeply about how I presented myself, on and off the court, so I made sure that my hair looked nice, and I wore makeup and checked to make sure my clothes had no wrinkles for these occasions. Otherwise I was always in tennis gear. Peachy, along with a couple of the other older women who were unofficial den mothers on the tour, instructed me in the proper dress codes for various events, always making sure my look was on point. I was nowhere near as refined in my appearance as I am today, but I definitely was more polished than most of my peers.

The other women on the tour who guided me who are no longer with us were Lee Jackson, a retired official who made sure we arrived on time for sponsor events and gave us tips on etiquette, and senior tour supervisor Georgina Clark, a stern fairy godmother from the United Kingdom who took no flak from us. These women cared about us as individuals and helped us manage the day-to-day responsibilities, from when to go to bed to mitigate jet lag while not cutting into our training time, to how to handle our expenses. It's not that they were officially responsible for guiding us. It was something they chose to do because they truly cared about our development as young ladies. This was especially important for rookie players like me. When you're running in circles with players who've already made a couple of million and they're buying a $400 sweater, you need to be fine with your $25 top. Peachy, Lee, and Georgina made it okay. They kept me grounded as I internalized these lessons and came into my own.

Having experienced every facet of the tennis world, I've come to appreciate what Peachy has done for our sport on a more holistic level. I've also learned from her gentle, unassuming manner. Peachy, who was raised in Charleston, West Virginia, is another classic Southern lady. But underneath that gracious demeanor is a woman fiercely determined to drive change. Watching her has taught me a lot about the velvet glove approach.

Over the years, we've grown especially close as we worked together on various boards, attending a multitude of meetings and, even better, long dinners over a bottle of wine. There are few things I enjoy more than sitting down with Peachy reminiscing over those pioneering days of the WTA, working together to take women's tennis to the next level, and the next.

These coaches and mentors helped me to grow both as a player and as a person. But my teachers didn't come exclusively from the tennis world. I've spent my entire career learning all that I could from the places I have been and the people I've met, regardless of their official roles and titles. Accept the gifts; soak up all that you can from the circumstances in which you find yourself, whatever they may be. My development as an organizational leader could not have happened without the countless conversations I've had with individuals like Jay F. Hans, who became my agent and accountant, a couple years after I joined the professional tour.

Jay got into representing tennis players by accident. The former marine and Vietnam veteran was representing a few NBA players and doing some accounting for them when an attorney friend of his convinced him to invest in the Virginia Slims of Nashville. In

1991, I was awarded a wild card to play in the main draw and lost in the finals to Sabine Appelmans of Belgium. Jay was impressed by my play and attitude, and over time we developed a friendship.

I already had an agent. But there was something about this burly blond giant and his no-BS way of communicating that my father trusted. The fact that he was doing taxes for many of the other players on the pro circuit was also assuring. Dad was always wary about the way young players might be taken advantage of when they turned professional and thinking ahead to various ways he could protect me when he couldn't be there. He'd heard many stories of "managers" ripping off athletes in all sports. "I need you to look out for my daughter," my father said. When I parted company with my initial agent, Jay became my go-to guy for advice on everything from finance to career development.

It was through Jay that I got to know my doubles partner, Manon "Bollie" Bollegraf. Wherever we were on the tour, whether it was London, Paris, or Zurich, Jay would invite me and the other players he represented out to dinner. Jay would hold court, talking about life, doling out career advice, and taking the edge off the day. Beyond being good with numbers, Jay's insights and war stories were a healthy reminder of the world beyond the tennis bubble we were living in. Jay made a point of getting me to think about the next phase of my career, encouraging me as I mulled the possibility of joining the Tennis Channel and cheering me on as I became more actively involved on various tennis boards.

"The game is the short part of life," he told me. "The rest of it, God willing, is much longer, so think about what you want to do with it."

Jay will always be that trusted sounding board as I venture farther outside the tennis world into more corporate and private company board positions. He sees himself as my protector, always urging caution, raising questions about potential liability, the specific dynamics of a situation, and who is making the decisions. He makes sure I do the necessary due diligence, drilling down into an organization's financials and history. Even in nonprofits, there's always the possibility of misuse of funds.

"Don't get caught up in the bright lights; get down to the meat of it," he told me one day as I was mulling options. "Because you're not there for the day-to-day operations you have to be sure. I'm an old guy, and if it smells it stinks."

The farther I've gone in the third phase of my career, the more I've benefited from a kind of reciprocal mentoring, where we take turns learning from each other, leveraging our respective life and work experiences to round out our own. I've done this with dozens of colleagues at the USTA, including Stacey Allaster, former WTA CEO and current USTA Chief Executive of Professional Tennis, and US Open Tournament Director and one of my men.

Stacey, a native of Canada, took a slightly different route to sports leadership. She got her first job at twelve, cleaning the clay courts of her local tennis club. After earning her law degree, she rose through the ranks of Canadian tennis to become tournament director of the Rogers Cup before crossing the border to join the WTA, which she built into a $1 billion organization and drove its expansion into Asia.

Stacey, who was listed as one of *Forbes* magazine's top women in sports, joined me a few years into one of the most ambitious upgrades in the USTA's history—a complete $600 million rebuild of the USTA Billie Jean King National Tennis Center that was completed on time and on budget. A consummate planner with a sharp eye for detail, one of Stacey's roles was to make sure every aspect of the players' well-being was at the forefront of renovations to the player service area, café, offices, nursery, warm-up areas, and many other facilities.

She joined the USTA just as we were focusing on the rebuild of player areas, so it was perfect timing. Having run the WTA Tour, Stacey knew how to execute a plan to fill the cracks of the players' needs regarding which we were not fully up to par. Entourages were growing, as were the number of children being brought to events, so quarters were getting cramped. But different stakeholders had different agendas for how best to utilize our landlocked facilities. Stacey understood how to negotiate the competing interests, helping us to find creative solutions, such as relocating and repurposing the broadcast studios and other spaces previously reserved for other functions. By working with Danny Zausner, chief operating officer of the USTA Billie Jean King National Tennis Center, who was responsible for the entire facility, the result was expanded facilities in which players, coaches, agents, and physiotherapists could feel comfortable. Stacey did this while overseeing tournament operations alongside her then tournament director, David Brewer, attending to the needs of tournament officials, and keeping Gordon Smith and me informed on managing the twenty-four-hour planning cycle of the US Open. All the while, she was grace under immense pressure. Her calm amid the chaos gave me confidence.

We fed off each other's energy, drive, and humor. When we first stepped into our roles at the USTA, the photos of past presidents on the wall at our headquarters in White Plains numbered seventy men and three women (my picture was not up there yet). Including Stacey, the organization added one more female on its senior management team, raising its total from two to three out of eleven. When one of the board members took the opportunity to talk about the "growth" in numbers of senior women on staff, Stacey couldn't help herself:

"That's a pretty low bar! This is nothing to celebrate, ladies and gentlemen."

We looked at each other, smiled, and winked.

You can learn from someone whose style, perspective, and background are completely different from your own. There's much wisdom to be gained from seeing the view from a whole other vantage point. Again, there is so much we learn in our sport that applies to business and nonprofit leadership. Tennis is a game of strategy. Each point must be orchestrated, setting up for the winning shot. That requires constant agility, knowing when to move forward or retreat, how far to push, or when to simply watch and wait. Stacey's experience in the C-suite allows me to lean on her when I have questions or concerns. She is all about positive change for women, and she is motivated by knowing that she is able to bring other women along, giving them the tools needed to succeed. I had always admired her from afar, but now I had a chance to work with her and learn from her.

It's exactly parallel to the tactical execution required to achieve an organization's goals. Hurtling full speed into a situation without taking the time to assess it from all angles can lead to setbacks. Over the years, I've learned to blend my aggressive style with patient observation, humbly taking the time to learn and bring others along, especially in large, complex, and team-based projects. It's the only way to lead effectively.

Former USTA CEO and executive director Gordon Smith and I served on the board together for a few years before he was hired as the executive director, and I assumed the leadership role, but we weren't exactly close. In fact, Gordon was among the old guard who initially had his doubts when, in 1998, the national governing body of tennis and other sports, USOPC, made it mandatory to bring a percentage of former top-seeded players to serve on the board. It was a controversial decision.

"No one likes being told who you should have on the board," he later shared with me.

We didn't always agree, but when you respect someone you find a way to make it work. Gordon was exactly the kind of person I needed in my orbit—a mentor with a specific set of skills and experience—who could help me to flourish in this new arena of organizational leadership. What we had was a kind of mutual mentorship. As he got to know me, he realized the value of having a former player's perspective. As I got to know Gordon, I learned the value of having someone on my side whose charisma and diplomatic skills were world class.

A trained litigator and born politician, Gordon never stated a problem as a problem. As a direct, tackle-it-head-on kind of girl, I thought he was bulls*#%ing me. Then I realized that was just his way of managing people. He wanted to create a calm climate. They could trust that there wasn't going to be any drama when the inevitable mistakes were made, such as putting the wrong information in a press release. He preferred that they spoke up if there was an issue rather than try to fix it by themselves or cover up, digging themselves down a deeper hole. He wanted to have his finger on every aspect of the business.

We felt the benefit of Gordon's calm leadership style when the USTA went through its digital transformation, slowly getting caught up to the twenty-first century by putting our entire business online. This involved upgrading our Tournament Data Management (TDM) system, which handles all entries into events. Back in the day, players had to register events manually. Match schedules would not be announced until the night before, making it difficult for players and their families to plan accordingly. But a TDM system handles all of these functions digitally. Players can register online for an event, seeing the draws well in advance so that everyone knows who is playing whom over a tournament weekend. It lets you know instantly about rain delays and match rescheduling. But when our IT team began the process of enhancing the entire digital system, they had to delink the TDM, keeping it off-line so that it could continue to operate. This "upgrade" shut down the entire system, resulting in dozens of tournaments going off-line in the middle of July, during one of the busiest weekends of the tennis season. USTA staff across the country had to put in long hours writing everything out manually, contacting hundreds of players and parents by phone to let them know what was happening.

We survived it, but not without a few scrapes and bruises. It was the kind of tense situation that could have caused members of our staff to have a meltdown, but Gordon's unruffled demeanor kept everyone as calm as possible under the circumstances, helping to keep things moving along until the massive technical glitch could be corrected. There were a lot of phone calls and emails flying around with not the nicest tones behind them, but Gordon and his team managed the chaos as best they could. It was an embarrassing debacle, but we learned a lesson.

Gordon also taught me patience. Organizational change doesn't happen overnight. Pushing too hard, too fast can backfire. When I became president and chair, I had a long list of improvements I wanted to make, from greater inclusion at the administrative level to increased grassroots investment to make the game more accessible to Latino Americans. Gordon collaborated with me on the best timing and strategy for getting board approval on these and other major investments, including the USTA National Tennis

Campus. However, over time, I realized that his approach, being charismatic, was his weapon to get what he wanted accomplished—and that was not always the same as what I wanted to accomplish. Recognizing this is extremely important as a leader because you can find yourself making bad decisions if you're not careful. Although Gordon had evolved in his role and spoke about equality for gender and race, I felt that he was still one of the "good ole boys," and it took a lot of hard work to change his views on many topics. If only Gordon could have delegated without fear of losing control and not being given credit, he would have been an even more efficient team leader than he was.

Whatever my vision, whatever promises I wanted upheld, I quickly came to realize that I had to nudge Gordon a little harder to dive into the weeds and execute with his team. He covered the day-to-day. Of course, it was also incumbent on me to follow up, stay informed, and do my own research. Delegating doesn't mean giving up the reins entirely because things could easily get twisted or sometimes overlooked.

When I first joined the board of directors, Gordon's predecessor Lee Hamilton's title was chief operating officer and executive director. Gordon succeeded Lee, but over time, that title no longer adequately reflected his role or that position. He'd done so much during his tenure, including improving the internal culture of the organization and bringing together different departments to work more cohesively and productively. In addition to leading our ambitious and complex digital overhaul, he led his team as they launched a new youth initiative (Net Generation) and initiated the introduction of electronic line calling, livestreaming from all of the US Open courts, as well as the doubling of revenues and record increases in US Open attendance.

When I became the chairman, CEO, and president in 2015, I

recognized that those three titles didn't belong together. I also felt that the tenure of this position of two years was too short. How could anyone accomplish their goals in just two years? We had reviewed this about ten years before I assumed the role, and the two years was upheld. It was time to review it again. I put together a Governance Task Force, consisting of leaders from the sections, board, and senior management who had law backgrounds. Their charge was to seek support for a three-year term for the chairman and president and to dive in and review our governance on many levels, including the title of CEO. Remember the USTA is a 501(c)3 and does not operate like a for-profit business. After a year of review, their recommendation was to keep the tenure at two years to give more opportunities to the hard working volunteers to lead the organization, but to transfer the title of CEO to Gordon as the executive director and the chief of staff.

He deserved the title, and the organization needed more modern governance, with a more clearly defined role for the chairman and the CEO. I would remain chairman and president, but Gordon's new CEO title would give more clarity to our functions when we were meeting with outside sponsors, for example. There were several business actions that needed the signature of the CEO, but as a volunteer, my signature wasn't legally viable. That signatory function fell to Gordon as the COO and executive director, which was confusing to those outside the organization; so for purely practical purposes the titles were not working.

My role as president and chairman was to be a visionary and motivator, to lead and give guidance to the chief operating officer and executive director and other senior staff. But the staff worked for Gordon, who reported to me and the board of directors.

My leadership role was more than a nine-to-five because that's how much I was entrenched—but no leader works only eight hours.

However, much of my time was on the road, domestically and internationally, meeting with other leaders of the sport, particularly on the professional tour at the Grand Slam events and other major tournaments, discussing critical aspects of our business. I was also the face and voice representing Team USA at the Davis and Fed Cup competitions and the 2016 Olympics, supporting our players and staff. Overall, this was a very intense and important responsibility.

Many of the old-timers on previous boards were dead against the title change. But I personally could not accept the status quo. Besides, CEO was a title I didn't need, certainly not as a volunteer. It mattered more to many of my predecessors, who were retired from their careers or had taken a leave of absence during their term. Knowing that after my tenure there would be more young leaders in the pipeline, many coming from corporate America, was another reason why I didn't feel the CEO title should continue. It was too difficult to explain that we were volunteers at a nonprofit organization with $1 billion in assets. Even though it's a volunteer position and not nine to five, people naturally assumed it was a more hands-on, operational role. In terms of time and interests, they perceived it as a potential conflict with the USTA leader's day job.

Gordon didn't ask for the CEO title. But I was determined to get buy-in from the rest of the board. If we wanted to professionalize the organization and attract more volunteers from the corporate world, we needed more clarity and good governance. In a corporation, CEO would have naturally been Gordon's title. It was a matter of matching his function with the recognition he deserved. After all, the CEO's job was to enact the policies set by the chairman and the board, which was what Gordon had been doing all along.

Deep in his heart, Gordon was happy. This time, he'd learned a thing or two from me.

8

HOME COURT

It's not our differences that divide us. It is our inability
to recognize, accept, and celebrate those differences.

—AUDRE LORDE

Maybe it's because I did so well on grass courts that I have a special fondness for Wimbledon, the only tournament in my history of playing where I ironed my skirt and shirt. It's also the only tournament where men and women players must wear all white. Everything, even your underwear, must be whiter than white, so that there's no risk of anything becoming visible on the courts that is not as pristine as newly fallen snow. The reasoning behind this tradition is that athletes don't stand out for anything more than their brilliant playing although, being black, I stood out everywhere.

But I must confess that in 1994 I did break this rule. As a fan of bright colors, particularly purple, the color of Northwestern, sticking to pure white was tough for me. I had a sponsor, Fancy Pants, which designed colorful and artistic tennis panties. I had the audacity to wear a pair under my white tennis skirt for one of my Wimbledon singles matches. It was a blustery day, which meant gusts of wind frequently blew up my chiffon tennis skirt while I was playing on one of the front courts by the Royal Box Terrace, making me a prime target for photographers.

After the match, I was summoned by the reporters. This was unusual, because press conferences are reserved for the winner, so I was intrigued to learn what this was all about. Standing in front of the paparazzi, someone asked me about my tennis panties, or what the British call "knickers." It seems they were much more interested in what was under my skirt than in the outcome of the match.

"Your colorful clothing caused a bit of a stir on Court Five," they informed me. "Can you tell us about that?"

I thought, *Seriously?* Then I explained they were provided by my sponsor.

"We all thought they were very attractive," one of the male reporters piped up.

"Thanks a lot!" I replied. "You saw my bottom a lot today, huh. No Peeping Toms or anything!"

The next morning my hot-pink, black, and green behind made all the London newspapers and tabloids, one of which printed WIMBLEBUM in big bold letters on its front page. When I read that headline, I burst out laughing. I thought it was hilarious, as did everyone else apart from the officials. Other than giving me some side-eye, they said nothing directly to me about the incident. But, since that day, players wouldn't even make it past the locker room exit without a knicker inspection. I don't think anyone has seen colored tennis undies on those hallowed courts since, until recent years when Serena donned orange undies to match her orange soles on her shoes, which was ultimately banned.

In the USTA President's Suite, we have our own strict dress code. Not as strict as Wimbledon, where male guests trying to enter the Royal Box will be turned away if they're not wearing a jacket, tie, and socks. But we're explicit about our no jeans, shorts, or sneakers rule. It's spelled out on the invitation in boldface type. Some people do miss the memo. Their assistants either failed to inform them or they simply, innocently, forgot or ignored the info. Most apologize profusely before running next door to the Ralph Lauren store to pick up a pair of chinos or borrow a pair of slacks, a dress, or shoes from our loaner stash. Most.

One glaring exception was a rich New York City businessman who, during the second night of the US Open in 2016, threw a hissy fit when we attempted to turn away one of his guests, a male model from Argentina, for wearing jeans. Clearly, he felt his money entitled him to a pass, regardless of the rules. So when he didn't get his way, he made a scene, berating and insulting my suite staff.

"Those jeans cost six hundred dollars," he told Cathy Politi, who does the job of a dozen air traffic controllers keeping the suite running smoothly, shift after shift. She makes sure our VIP guests are fed, watered, and comfortably enjoying the session of matches they've been invited to view, ushering them out gently to prepare the suite for the next wave. The vast majority are courteous and more than happy to abide by our house rules. But not this gentleman.

"My friend is pairing them with a chic Prada blazer," he told her. "His outfit is worth more than anything you people are wearing."

On and on he went with his breathtaking rudeness, swearing, badgering, and belittling our staff. I was shocked to find out later that he had his eighty-six-year-old mother with him.

I didn't witness the initial exchange. I was outside, sitting with guests where we'd been watching an intense battle rage between Andy Murray and Nick Kyrgios, when Cathy came and whispered in my ear that someone was making a scene. During a break between games I excused myself and darted up the stairs to handle the situation. When I got there, our front desk staff were visibly shaken. No way was I going to allow my team to be disrespected. It was inexcusable. If someone had enough of a problem with our rules to become verbally abusive, I would gladly take that heat.

"I'm very sorry sir, but rules are rules," I explained. "We enforce a strict no-denim dress code."

"Do you know who I am?" the outraged donor asked me.

While his jeans-clad friend stood out in the hallway, this "gentleman" then attempted to make his case to me.

"I'm wearing gray jeans," he told me. "There are people here wearing dark jeans. . . ."

While I wasn't about to do a swatch test or spend the rest of the evening litigating over what pants did or did not qualify as jeans wear, it was clear no one else in the suite was wearing blue denims, however expensive they supposedly were. But when he realized I wasn't buying it, he changed tack, proceeding to educate me on the details of his generosity as a donor, loudly informing me he'd just cut a check for the City Parks Foundation tennis programs an hour earlier.

"And I've given five hundred thousand dollars over the years, but you won't get another dime if you don't do the right thing and let my friend into the suite!"

My response to these situations, which fortunately are rare, is to smile and, in a calm, even tone, stand my ground. Once again, I patiently explained the dress code to this gentleman, and the fact that we make no exceptions, even when they are celebrities or heads of state. Changing our position would be unfair to more than four hundred people who pass through the suite each day with no problem respecting the rules.

Enforcing the code can lead to some uncomfortable moments, but the way certain guests respond is like night and day. I had to personally ask Alex Rodriguez (A-Rod) to leave because he was wearing jeans. He was the guest of a guest, so he didn't know the rules, but he was extremely polite about it, immediately getting up to leave.

"Please don't get up yet," I told him. "Just wait until the end of the set and enjoy a bit more tennis before you make your exit." (It was 5-4 in the set.)

The set ended up going into a tiebreaker, which gave him more time. A man of pure class, he came over to thank me for my hospitality.

"I hope to see you later in the week, and I promise I won't be wearing jeans!"

Another year, we unfortunately had to turn away Jerry Seinfeld for wearing denim. He made a stink about it, but we arranged for him to watch the match from one of our sponsors' suites, and he went his way. We regretfully had to send out Al Roker and his son, who was dressed beautifully in a jacket, crisp shirt, and tie but unfortunately had on a pair of immaculate white sneakers. Mortified, the Roker men sincerely apologized, then went off in search of appropriate footwear.

Eventually, this irate guest and his denim-clad party went their way. The next day I got to read his inaccurate version of events on Page Six:

"She cared more about faded jeans than giving money to kids," sighed the scene-making gentleman, adding he'll pull support from the event after being asked to leave. "Celebrities come in flip-flops . . . I couldn't get over her rudeness. I don't have a problem with the USTA. I have a problem with this woman."

The following year, he told the *New York Post* he planned to rent his own suite where "only people in faded jeans are permitted." I haven't seen him since.

Wherever you are in the world, be it your home court or a land where the culture is completely different from your own, it's critical to know how to represent. Being respectful, polished, and aware of the space I am entering has brought me the level of access and credibility necessary to make critical connections with people of influence. I first learned how to represent from my parents, who were both immaculate in their self-presentation. Mom never went anywhere without putting on lipstick and a pair of three-inch heels, whether that was to her classroom, the doctor for appointments, or church on Sundays. I try to live up to her standards because it shows I care, which earns me respect in turn. Missing that point by walking into any arena with a sense of entitlement or indifference to the rules or cultural sensitivities of others does just the opposite, creating a lasting negative impression that may hurt your game both professionally and personally.

For an example of how to embody that winning combination of humility and graciousness, look no further than Roger Federer. Besides being ridiculously attractive, he has the most welcoming manner of any of the tennis professionals I've ever met. Whenever I bump into him at a tournament he doesn't just say hi in passing. He stops, embraces me with a European kiss on both cheeks, then takes the trouble to ask me a specific question about my family, remembering details from conversations we had months earlier. Much as I'd like to think it's because I'm special, Roger doesn't just do that with me. At a recent event, I watched him out of the corner of my eye. He was deeply engaged in a conversation with a fan. He didn't take his eyes off that person the entire time they were chatting.

Roger is an extremely busy man, with countless people and things competing for his attention at any given moment, but that didn't matter. He insisted on being present. His press person was getting visibly impatient, looking at his watch and trying to move things along, but Roger was intent on seeing out the conversation until there was a natural moment to excuse himself and move on to another friend. Not once did he appear distracted. He's one of those people who makes you feel like the most important person in the room. It's no wonder he's been wildly successful beyond his record-breaking Grand Slam tournament titles. Roger has built his personal brand on kindness, generosity, and class.

Learning how to navigate these different worlds like a professional led me to become a player ambassador, an international representative of the sport through board membership and media and an executive of global tennis organizations, the chairperson of the Gender Equality in Tennis Committee, and, most recently, a financial company board member. My "finishing school" of life experiences in the tennis world have given me the confidence to chat in the Royal Box with British princesses at Wimbledon and enabled me to charm, persuade, and negotiate for better conditions for female players with leaders of all tennis associations. It's also helped me to hold my own in conversations with Fortune 500 CEOs and other big donors to the grassroots tennis programs that are near to my heart. When I am the only person in the room who looks like me, I am intentional about how I engage with others, knowing that what I say and how I say it will likely stand out more than others. It's why having the social skills and etiquette has been a huge asset.

Every Grand Slam tournament has its set of rules and customs, influenced by the culture of the city where it's located. Over the years, I've made plenty of friends in the Land Down Under, which I consider to be my second home. In Melbourne, the Australian Open is more relaxed and a more casual dress code is welcomed, accessorized with Aperol spritzes or cold beers for the searing-hot summer days. At twenty rows up, it's a hike to get up from the stands to the "O" (formerly known as the President's Reserve) and can take an entire changeover to make the trip. You may prefer to stay out on the grounds, where they have beer and wine gardens along with live music.

In Paris, Roland-Garros is a chic but sometimes congested experience. The Court Philippe-Chatrier (center court) fans are dressed fashionably, with many wearing fedoras paired with classy sunglasses. Champagne flows, as well as Perrier. The grounds are confined so you can easily get shoved and bumped, so it's important not to text while walking. Until 2018, my final year as president, the President's Tribune, their version of our President's Suite, is just a foyer stocked with petits fours and drinks. Ice is a delicacy, so on the hottest days, don't expect to get a cold drink. One ice cube has to suffice, two if you smile. Drinks aren't allowed in their distinguished seating area, so you'd better hydrate before going to your seats. However, the stadium has just undergone major renovations, adding a roof and modernizing the President's Tribune.

I love the proximity at Wimbledon on the outside courts, where you can sit just a few feet from the players. But, on Center Court, only the sideline seats give you that same experience. Most men wear a suit and tie, while women often wear sundresses. You can wear a panama hat, but nothing too high or oversize that risks blocking someone's view from behind. The crowd seldom cheers in the middle of rallies, respecting the rule of total silence until the end of each point. They stay in their seats until after the third game of a set or, when the players change sides, every two games, even if all that alcohol leaves them busting for a bathroom break. The only creature allowed to move is Rufus, the hawk that's been trained to fly around the stadium keeping the other birds away.

There are also royal protocols. Back in my playing days, we had to bow or curtsy to members of the royal family upon entering or leaving the Center Court. The custom ended in 2003, when the Duke of Kent, president of the All England Lawn Tennis and Croquet Club, deemed it too anachronistic. Now players are only expected to do so for the Queen or the Prince of Wales, although they rarely make an appearance. If you happen to be in the box with members of the royal family, it's a whole other set of rules: you call them "Your Royal Highness," then sir or ma'am unless it's the Queen, who is "Your Majesty." You only shake hands when one of them reaches out to you first, and you sure don't give them a hug!

Roberta momentarily forgot this protocol when she greeted the Duchess of Gloucester, although the Brits tend to be understanding of well-intentioned Americans and our less formal ways.

"It's so good to see you again!" Roberta gushed, as the Duchess came out to greet us on the terrace of the Royal Box.

The rest was like a slow-motion movie scene as my girl stretched out her arms to embrace the blue blood. In my head, I screamed, *Noooooo!* But it was too late. The Duchess had no choice but to return Roberta's friendly and overly familiar hug.

Roberta and I were there with Dave Haggerty, my predecessor at the USTA, and his partner Liz Leary, as well as then president of Tennis Australia, Steve Healy and his wife, Cathy. Being from a Commonwealth country, the Australians knew better. But before we could politely intervene, Liz followed Roberta's lead and went in for her own hug.

"I am so sorry!" Cathy, Steve, and I said in unison.

"Why, what happened?" Roberta asked.

She was mortified when we explained, especially after she recalled the trouble Michelle Obama got into when she gave the Queen a friendly over-the-shoulder squeeze a year earlier.

"Not to worry," the Duchess reassured us. "Protocol always gets in the way of friendship."

The younger, newer members, like Kate Middleton, the Duchess of Cambridge, and Meghan Markle, the Duchess of Sussex, also happen to be especially warm and down to earth. The Duchess of Sussex and I shared our love and admiration for our mutual friend, Serena, and our experience attending Northwestern University.

I didn't mind adhering to these traditions, anachronistic or not. Both as a player and as a representative of the sport, I wanted to be in sync with the surroundings: the immaculate grass courts, the ivy, the flowers, the tea and scones with strawberries and cream . . . It was an oasis of perfection. Nothing was out of place, not even the grains of grass. Fitting in and being more pristine in the way I presented myself made me walk with my chest stuck out. I had a sense of pride in being there that made me that much more confident as a player. In effect, I adapted myself to a new culture and grew.

It's an altogether different type of energy at the US Open, which is of course my favorite of all the Grand Slam tournaments. Tennis legend Jimmy Connors put it best: "New Yorkers love it when you spill your guts out there. Spill your guts at Wimbledon and they make you stop and clean it up." This tournament is so New York: boisterous and intense. Fans wear their emotions on their sleeves, and they're not shy about sharing. I always felt the pressure here, because it was where more of my friends and family members would come to watch me play. The fact that the matches can go on well into the night under the bright lights of the stadium, and that we players compete for the largest purse of all the major tournaments, also stokes the tension.

But what really makes the event crackle is the interaction between players and fans. Russian player Daniil Medvedev has described how he feeds off that energy, but not in a way you'd expect. When he knows the crowd is against him, he grows defiant and hits harder.

Daniil became the favorite villain of the 2019 US Open when he started playing his third-round match against Spaniard Feliciano López. During the first set, he snatched a towel out of the hands of one of the ball boys, prompting the New York fans to boo him. His response? The middle finger, which happened to have been shown on a large monitor. He went on to win the match in four sets, then "thanked" the crowd for their jeers.

"Thank you all, because your energy tonight gave me the win," he told them. "The more you do this, the more I will win for you guys."

From that point on, cockiness mixed with rebellion became Daniil's thing. He was booed again as he walked onto the court for his fourth-round match with German player Dominik Koepfer. When he won, he encouraged the fans to boo once more, and when they came through for him, he did a little dance to celebrate.

"I was painful in my abductor before the match. I thought, I'm not going to play . . . And you guys, being against me, you gave me so much energy to win. Thank you!"

He continued to use that energy to propel him all the way to the Men's Singles finals and a close match, where he nearly toppled Rafael Nadal, coming back from two sets down and losing in five. I've got to hand it to the guy for turning a negative into a positive.

Daniil's "special" relationship with New York fans notwith-standing, there are still rules. The crowd must keep quiet while the ball is in play; well, they are supposed to but there is a constant buzz in Arthur Ashe Stadium. The crowd is allowed to go crazy at the end of a point. They should also refrain from heckling a player, which could result in being ejected from the match. Mockery in any form, including making fun of grunts as players serve and re-turn shots, is frowned upon. But stuff happens with thousands of the world's most passionate fans.

At the 1990 US Open, during an early-round doubles match I was playing against Martina Navratilova and Gigi Fernández, one of those fans happened to be my brother, Maurice, or "Recee" as we call him. Each time I won a point, he would scream, "Yay!" That was embarrassing enough, but at least it was inside the lines of US Open etiquette. When I won the first set my brother, who is loud with a huge personality, leaped out of his seat, shouting, "That's my sister! That's my sister!"

I'm more than happy for Maurice to light up a room just by being his fabulous self, but when Martina and Gigi were winning in the second set, he started booing. He did not know the rules and was so overly excited I doubt he'd have cared. Martina asked for him to be removed from the stands, but he paid her no mind, continuing to jeer at her every time she scored a point. Of course, I was mortified. I kept looking up at my brother in the stands and giving him what he calls "that look"—to no avail. He eventually calmed down as we proceeded to lose the match. Thank God we weren't at Wimbledon.

It's not always easy to navigate cultures completely different from your own. Throughout my own development as a global am-bassador there have been all kinds of rules, written and unwritten, I've had to learn, as well as varying points of view I've needed to familiarize myself with, whether I agreed or not. Even within the US you're going to come across folks with polar-opposite opinions on certain issues, especially in today's divisive political climate. The trick is to give people the benefit of the doubt. The older generation doesn't necessarily grasp the concept of alternate personal pronouns, for example. Language and cultural sensitivities have evolved rapidly over the last few years. Depending on their demographic and life experience, people are still catching up to what is considered acceptable. Don't go looking for offense where none was necessarily intended. Assume the person you are interacting with means well, but don't overlook your intuition.

That was my challenge when traveling as vice president of the International Tennis Federation. These are the moments when it behooves you to meet people where they are. At a recent ITF Annual General Meeting in Orlando, I was approached by a delegate from Saudi Arabia who asked me if he could sit in on our Gender Equality Committee. I had just given a presentation on our goals and it struck a chord with him.

"I applaud what you are doing and hope to do more to advocate for female leadership, but in my country, women only just got the right to drive," he told me. "We're not going to move the needle anytime soon, but I'd love to watch and learn." I had just unveiled our "Advantage All" campaign, signifying that tennis is an equal opportunity sport. We wanted to highlight the importance of getting women into leadership roles, whether they were on a committee, commission, or board. We also wanted to see more women as coaches, in development, or as officials. These are areas where the USTA and a few other nations' tennis associations have been leaders but not so much for the rest of the world. The ITF wants women to feel empowered, valued, have a voice, be a part of the culture change, and have balance in decision making.

That evening he came to our annual gala, where delegates dress in their native costume. It's one of my favorite events of the year, festive, and full of color, where people come in their dashikis, saris, *salwar kameez*es, kanzus, and *kanga*s. But had I judged this gentleman solely on the long, white Saudi dishdasha robes he wore to this event, or any other guest in native dress, instead of the content of his character, it would have been my loss.

In my global leadership role, the ability to accept differences has enabled me to excel. I'm okay with the fact that in conservative cultures you dress appropriately, cover your shoulders, and make sure you cover up out of respect for their customs. I don't take it personally when I am running a meeting in a boardroom full of men whose wives weren't even allowed to go out shopping by themselves. Instead, I focus on moving the needle ever so slightly forward on issues I care deeply about, like gender equality. Not every political mindset or experience is the same, but I always try to find common ground, whatever country someone is from.

To that end, it helps to get to know others as individuals rather than make assumptions based on their national, cultural, or racial identity. While it was still frowned upon by other African American athletes, I happily shared the court with white South African players—nine in all. Just because their government practiced apartheid didn't mean they were personally racist or in favor of segregation. If anything, these players' willingness to be seen playing alongside me on the television screens of Johannesburg, Cape Town, or Soweto could only further the cause of racial equality back in their home countries; but of course, that was only if we played on a major court. I was right about their intentions in every case but one: an Afrikaner woman who shall remain nameless. It was clear on a number of occasions that she didn't care for black people by comments that she would make or how she treated essential workers, demanding service or just being rude. That she wanted to win some partnership with one was the exception. I am friends with the other players to this day. I felt that my example of trying to break down the color barrier for these players was critical. I had mad respect for Arthur Ashe when he went to South Africa in 1973 using tennis to inspire social change. The bricks had been laid by one of my heroes, and I felt by playing with them that people watching and cheering would see that we were human, using sport to address racial barriers in South Africa.

You don't necessarily have to agree, but you will evolve when you listen, put yourself in another's shoes, and find a medium where you can coexist. Practicing this kind of tolerance and understanding is a part of doing business in the global economy, whatever the industry. From the US to Europe, some political leaders may be trying to turn inward, tearing up decades-old trade agreements and closing borders to immigrants. But successful businesses keep moving forward into the rest of the world, building and investing in regions and countries far beyond their headquarters. Their CEOs understand that it is in these less familiar cultures where the real opportunities for economic growth will continue to exist, despite what the politics of the day may say.

So, keep your own borders open. As you walk through all the arenas of the world, this expansive attitude will serve you well, opening you up for impactful lessons you might otherwise have missed. Representing by being the best version of yourself is as much about making other people feel at home as it is about bringing your home court advantage with you wherever you go.

This is especially true when you feel like you are the only one. Although it may seem lonely at times, you can't let an outward appearance determine who you are or what you stand for. That's why I am grateful to have a global platform. It's made me realize that we are never truly alone. Our differences can be very subtle in substance even though we are a world apart in demographics. Be confident in your knowledge, what you bring, and the singularity of your outward appearance, because when your experience and preparation are superior to others these qualities will rise above everything else.

Be proud of being different, being black, being a person of color, being a woman, being the only one. It empowers you to use your voice, your experience, your platform to make a difference and be a leader. It enables you to reach back, mentor another, and bring them forward. It inspires you to lead with a purpose and not be denied as the only one. By balancing your expertise, network, and environment, you have the ability to change the culture and motivate those around you to embrace people's differences and strengthen your business through diversity of thought.

Own your presence, own your knowledge, own your arena, own your voice.

9

RESPECT, REFLECT, AND REPRESENT

The mark of great sportsmen is not how good they are at their best but how good they are at their worst.

—MARTINA NAVRATILOVA

More than any other sport, whether it's football or soccer, where fans are more interested in an athlete's signature happy dance after scoring a goal or Gatorade victory dunks at the end of a game, all eyes are on the tennis players when they reach across the net at the end of a globally televised match, when the quality of that handshake is scrutinized. When you think about it, the importance of that moment in our sport makes sense when it's two people grueling it out on the court, sometimes for hours. The stakes are extremely high in an intensely personal battle of wills, so the ability to look your opponent in the eye and be just as gracious in victory as in defeat speaks volumes about a player's character.

Tennis broadcaster Tom Tebbutt put it perfectly when he wrote the following:

> The handshake, and the walk up to the handshake, is an integral part of a tennis match. It's the joy of victory and the agony of defeat, it's all the emotions of the competitors summed up in a final ritual gesture.

It's why I still cringe when I think about that postgame handshake between Sabine Lisicki and Agnieszka Radwańska after their Wimbledon semifinal match in 2013. Lisicki beat Radwańska 9-7 in the third set, winning her way into her first Grand Slam tournament final. The two played like true champions, then Radwańska stuck out her hand without even looking at her erstwhile opponent. It was a deeply human moment. I get it. Radwańska was heartbroken and just wanted to get out of there. You couldn't help but feel for her and the pain she must have felt in making that less-than-half-hearted gesture. But that ice-cold pressing of the flesh, and the lousy sportsmanship it signaled, was all people could talk about after the game. It completely eclipsed the great tennis the two women had played.

It's not the only time this has happened. Bad sportsmanship is endemic. Sometimes you catch players giving each other a slap of the hand. Doing it with that kind of attitude is almost as bad as no handshake at all. Then again, maybe not. In 2009, after the semifinal Women's Singles match in Eastbourne, England, Marion Bartoli refused to shake hands with her French compatriot Virginie Razzano over a previous spat. Apparently, she was annoyed that Razzano withdrew from the match with a muscle strain in her leg, while trailing 6-4, 1-0. Razzano later told a French newspaper that her compatriot often made a fuss over minor injuries to distract her opponent.

At the Australian Open in 2012, Czech player Tomáš Berdych refused to shake the hand of Spain's Nicolás Almagro, believing he'd intentionally hit a ball at him while he was at the net. Almagro apologized for the way his running passing shot landed, but Berdych had turned his back and didn't see his opponent's apology. The crowd booed Berdych throughout his courtside interview, although he later acknowledged that he regretted snubbing his opponent, saying, "Maybe we both did some mistake. So it's even, and that's it."

At the Davis Cup that same year, after a grueling five-plus-hour match in the quarterfinals, Czech Republic's Radek Štěpánek called the winner, Serbian Janko Tipsarević, a nasty word, then gave him the middle finger during their "handshake." A heated exchange ensued, during which the Serb had to be restrained by his team captain. But at least he didn't punch him in the chest, like Karel Nováček did to Derrick Rostagno after losing the first round of the 1992 French Open. Nováček, upset with Rostagno for critiquing his match conduct just as they were meeting at the net for the handshake, lost it.

Of course, no one can forget John McEnroe's rants at umpires during his heyday. When an umpire at Wimbledon made a line call John didn't like, he screamed, "You cannot be serious!" It's got to be one of the best-known lines of all sports, and John seems proud of it. He's the rare exception to the rule that court rages can throw you off your game. A good venting somehow made John play better, although I wouldn't recommend it.

Worse, in some ways, is taking that anger out on yourself. After hitting an easy return into the net, Russian player Mikhail Youzhny hit his own head three times hard with the frame of his racket, until he bled. It was so bad he had to get bandaged by medics, although interestingly he went on to win the match. Maybe he literally knocked his head back into the game, but this probably wasn't the best course of action.

In those moments of intense competition, our passion can cause us to lose the better version of ourselves. It happens, yet acting out has somehow become more accepted by match audiences. You might even say they enjoy the spectacle of athletes behaving badly. Fans almost expect it, and when something untoward does happen during or after a match, the media will invariably focus on the incident much more than the caliber of playing.

The non-handshake is the latest iteration of this unfortunate trend. After losing in two sets at the 2019 US Open, Camila Giorgi's post-match handshake with Maria Sakkari was so brusque it almost looked like a shove. That summer was a bad one for player conduct in general. In Bucharest, Eugenie Bouchard refused the hand of Alexandra Dulgheru after a draw for their Fed Cup tie, and Guillermo García-López left Lukáš Rosol hanging, despite the fact that he was the victor, because he didn't appreciate Rosol's conduct during the match.

We've all been guilty of the half-hearted handshake after a frustrating match. I've done it, although it was never a reflection of my feelings toward my opponent. Whatever the reason, there's no place in sports for a slap, shove, or snub. It's never okay. We all need to hold ourselves to a much higher standard in the respect we give one another. By looking your counterpart, adversary, competitor, or challenger in the eye, you humanize them, and they humanize you. In this age of trolling and anonymous attacks over social media, we tend to forget that there are real people on the other side of that digital screen. Face-to-face interaction doesn't solve this, but these gestures of sportsmanship reminds us of how connected we all are, no matter which side of the net we're standing on.

With all the divisiveness that's going on in our politics and society, true sportsmanship has never mattered more. Which is why, when I was asked to accept the #HandshakeChallenge during Wimbledon in 2017, it got me thinking. I had one more year in my tenure as head of the USTA.

I was sitting on the players' terrace when I was approached by John-Laffnie de Jager, founder of the movement to have players shake hands *before* a match as well as after. A former South African player and current captain of the Davis Cup South Africa team, JL had been my Mixed Doubles partner at Wimbledon '98. JL was a gentle and affable giant at six foot four and embodied the spirit of sportsmanship. As a professional player, I got the sense that JL was just grateful to be on the court. He was generous as a doubles partner, competitive yet incredibly supportive of his teammates. JL was also a hell of a player—a champion in his own right. He's always been about giving back to his sport, having started his own tennis academy in Johannesburg to help create the next generation of South African champions. John's training philosophy is as much about instilling the right attitude in these kids as it is about giving them skills.

His #HandshakeChallenge, which had already been taken up by Tennis South Africa, is intended as a sign of mutual respect. Shaking hands before as well as after a match might seem like a small thing, but it sets a friendly tone. It says: "We might be about to try and kill each other on this court, but you are my worthy opponent." Done right, you smile, make eye contact, and give the hand of the person on the other side of the net a tight squeeze. Culturally, it's a whole thing in South Africa, where you can even find guides to different handshake styles, from a rigorous shake to a prolonged squeeze-pump action. No limp-fish handshakes for them. They do it like they mean it!

JL had already been successful in encouraging other sports in South Africa to take up the challenge, including rugby, soccer, and cricket, where opposing teams can almost seem like they are at war. The significance of the handshake in South African sports goes back to the days immediately following apartheid, when the country was overcoming racial segregation in all areas, particularly rugby, which was a white, elitist sport at the time.

Nelson Mandela famously shook hands with the blond-haired, blue-eyed Afrikaner captain of the Springbok team after they won the Rugby World Cup in 1995. By donning the green Springbok jersey and embracing Francois Pienaar on a global stage, Mandela was demonstrating to the world the possibility of reconciliation, forgiveness, and peace. He was showing us all how to bridge the racial divide beyond the rugby pitch. It was sportsmanship at its most powerful. Interestingly, several Springbok rugby captains and players were among the first to embrace the #Handshake-Challenge when JL introduced it.

Globally, other sports organizations have been implementing their own versions of the #HandshakeChallenge. In 2014, the International Federation of Association Football (FIFA), which represents football (or soccer) around the world, launched its "Handshake for Peace" campaign to commemorate the one-hundred-year anniversary of Christmas Day during World War I, when German and Allied soldiers laid down their weapons to play soccer in the middle of no-man's-land. Just for one day, the young men took a break from their mutual slaughter to come out of their trenches and kick a ball across the frozen earth. FIFA figured that if opposing sides of a war could shake hands and play football as equals, their players could do the same, setting a global example of peace and solidarity in the heat of their own battles on the football pitch.

"I believe every one of us is someone's hero," John explained when I asked him why he'd become so driven to promote shaking hands at the net. "Each one of us has the ability to change lives and inspire. Sport helps us to develop into everyday champions and you will always be remembered not only by your performance but more so by the way you played and conducted yourself in competition."

We implemented the #HandshakeChallenge at the USTA in 2018, during my final year. Seeing what was happening in our sport, I felt the need to do something that could help reintroduce the spirit of sportsmanship. Treating each other with respect and understanding, on and off the court, was long overdue. We started at the US Open Junior Championships, introducing the pre-match handshake slowly. I was cautious about implementing it all the way, especially among the professional players during the Grand Slam tournament. The last thing we wanted was for a player to fail to adapt to the new protocol and then have all the media attention focused on their lapse instead of the tennis itself.

It was gratifying to see how quickly some of the younger players took to it, although it's taking longer than we'd hoped for the pre-match gesture to become automatic for them. I don't see it happening at the professional Grand Slam tournament level anytime soon. Umpires haven't been insisting upon this new, unwritten code of conduct, and a handshake can be hard to mandate. A gesture of respect to your opponent must be authentic, from the heart.

In my first term as USTA leader I launched the Sportsmanship Initiative for high school tennis. We felt the behavior was getting out of hand not just with the players but also with parents and coaches. It was turning people off to staying in the sport for a lifetime. We came up with the Sportsmanship Oath, which went out to three hundred fifty thousand youth nationwide who play tennis recreationally, seasonally, and competitively. The idea was to get them to say it out loud before a match so that, by speaking it, they could feel it:

> I recognize that tennis is a sport that places responsibility for fair play on me. I promise to abide by the rules of the game, which require me to give the benefit of the doubt to my opponent. At all times I shall strive to compete with the true spirit of sportsmanship, recognizing that my behavior on the court is a direct reflection of my character. Whether matches end with my victory or defeat, I promise to conduct myself in a way that honors my opponents, my team, those who support me, and the game of tennis.

We need to start promoting the spirit of sportsmanship much earlier in an athlete's career by emphasizing it more in our youth programs and setting better examples as adults. We need to improve attitudes, then the appropriate gestures will come.

When I was nine, one of the coaches at a summer youth program on Chicago's South Side, Donna Yuritic, wrote a list of my achievements that season, but for all the matches I'd won, there was still one area where she wanted to see improvement.

"Katrina has a tendency to pout during matches."

I would get mad at myself for missing shots. But, left unchecked, a bad attitude could easily have seeped into other areas of my court conduct. I was lucky to have many adults around me who taught me better, beginning with Mom and Dad. Snubbing an opponent's outstretched hand would never have been tolerated. But, far from seeing parents and coaches reinforce sportsmanlike behavior, these days the internet is full of viral videos showing dads brawling with each other in front of their kids at Little League baseball games, and soccer moms screaming insults at each other from the sidelines. If that's what this next generation is seeing on a regular basis, we're cheating our kids out of an essential life skill.

Thankfully, tennis is the one sport where there are many more examples of sportsmanlike conduct than not. After all, it evolved from the palaces of nobility in England and France, so it's long been a game of polite customs—something you feel at the All England Lawn Tennis and Croquet Club at Wimbledon, where everyone is expected to mind their manners.

There's plenty about sportsmanlike behavior covered in the USTA's own eight-page, forty-six-item *The Code: The Players' Guide to Fair Play and the Unwritten Rules of Tennis*, including the principle that "Tennis is a game that requires cooperation and courtesy." Assuming the best of everyone's intentions is one of the underlying principles of the Code, which holds that "points played in good faith are counted" even if it's discovered later that the net was too high or there was service from the wrong side of the court. Corrective action can be taken only after the point is completed, to avoid disrupting the player's focus. When there is no umpire and players must make their own calls, the opponent must get the benefit of the doubt, and players should only call the shots that landed on their side of the net.

The code goes into extensive, painstaking detail for the sake of fairness. If, for example, one doubles team partner calls the ball out "and the other partner sees the ball good, the ball is good. It is more important to give opponents the benefit of the doubt than to avoid possibly hurting a partner's feelings. The tactful way to achieve this desired result is to tell a partner quietly of the mistake then let the partner concede the point." But that rarely happens.

The only item that may be going by the wayside is item 37,

which covers grunting. Yes, there's a rule about this. "A player should avoid grunting and making other loud noises. Grunting and other loud noises may bother not only opponents but also players on adjacent courts." Maybe so, but I don't think most people mind that Rafael Nadal grunts. Players and spectators are used to it. It's a part of his brand, which also includes some electrifying tennis as he leaves it all on the court. Of course, the fans are quick to abuse the female grunters. No one ever stops to think about the fact that a man's pitch is much deeper than a woman's, whose voice is perceived as annoying somehow. It's an unfair double standard.

There are certain players who don't need a code of conduct to be great sportsmen and sportswomen. They just are. Although there were a couple of ladies that weren't missed when they were away from competition due to injuries. The atmosphere was really quiet when they were away.

When Rafa topped Roger Federer, in Melbourne in 2009, he held back on his jubilation out of deference for the king's feelings. He knew how painful that loss was to Roger, and he chose not to rub it in. During a Hopman Cup match in 2016, American Jack Sock stunned his Australian opponent, Lleyton Hewitt, when Hewitt's first serve was called out, telling him, "That was in if you want to challenge it." Hewitt went with it and challenged the call, which was, in fact, in. The umpire, players, and spectators were all smiling and chuckling as they watched the instant replay. Hewitt won the match, but Jack Sock won the hearts of everyone who saw him willingly lose a point for the sake of doing the right thing.

Another true gentleman of the sport was Tim Smyczek during the 2015 Australian Open, where he played against Rafa in a grueling five-set match. In the fifth set at 6-5 up, Rafa had just broken Tim and was about to serve when someone in the crowd catcalled. Rafa lost the point and was about to serve again when Tim asked the umpire to give Rafa the first serve again. Rafa just managed to win the match but, as he said in a post-match interview, "not many people on the tour would have done what Tim did at 6-5 in the fifth set."

These players understand that showing respect for their opponent is also showing respect for the game. The next hour may be a battle, but they also know that the better each one of us is, the more compelling the tennis. Even though we were fending for ourselves and it was our livelihood, we respected one another. Off the court, our rivalries often transitioned into solid friendships. I've already mentioned the camaraderie between myself and some of the other women on the WTA Tour. Even though we were competing, we shared more things in common than not. Dealing with life on the road, we supported each other, some of us even sharing coaches, hotel rooms, and practice partners. We weren't all besties, exactly. I was much closer to some than to others. We were more like members of a highly competitive sports sorority.

Tournaments often involved hours of waiting for top-seeded players to finish so that we could have our turn on the practice courts. Whoever we were playing against at the Zurich Open, our friendship would resume as soon as the match was over. We would huddle together in the freezing-cold playing cards, teasing each other or dishing about our love lives. We often didn't get out of those drafty old stadiums until 2:00 a.m.

Being on tour isn't as glamorous as people might think. In fact, it's often a grind, forcing you to be away from your family and loved ones for weeks at a time. We got through it with laughter. Our training room was often called "the Kitchen" of the tour, because that's where players like Pam Shriver, Caroline Vis, Liz Smylie, Gabriela Sabatini, Arantxa Sánchez Vicario, or Rennae Stubbs would swap stories and rib each other. It seemed like Rennae always had something funny to say.

No one was immune from a little locker room teasing. We once played a practical joke on Kathleen Stroia, the current senior vice president of Sports Science at the WTA who was traveling with us as our tour physiotherapist. We'd somehow managed to convince her that two of the quietest girls on the tour were developing a fan club for her. We put monogramed items with the initials KS all over the training room: towels, Band-Aids, even tampons.

Today you can see that camaraderie shine through the post-match interactions of many of the top players. The traditional handshake has morphed into a kind of friendly arm wrestle or a "cupper," where the hand being shook is cupped by the other hand in a more personal and intimate gesture. Chrissie Evert and Martina Navratilova were known for the warmth of their net handshakes. Gabriela Sabatini was one of the first women players to do the arm-wrestle shake. Some players have even advanced to a sweaty hug at the net. Novak Djokovic is a hugger. He even swapped shirts with an opponent after a match, which is more of a European soccer tradition.

Despite the "ick" factor, I love the friendliness and intimacy of these variations on the post-match shake. As long as there is some kind of respectful acknowledgment of a match well played, I'm down with it. Andy Murray, ever the British gentleman, embraces the ritual of the handshake no matter what goes on during the match.

"There are certain things that happen that when the match is done doesn't necessarily influence how you feel about that person," he told a reporter after a contentious match with Roger Federer. "Things in the heat of the moment, you understand that things can be said."

Don't get me wrong. Andy doesn't always behave himself on the court, but he will always be courteous once the game is over. (I also love the fact that he is the first top male player to have a female coach.)

Showing respect for your opponent is tantamount to having respect for the game itself. If the person you are playing against didn't train and fight just as hard as you, the quality of the match would be diminished. You're not just battling each other to win, you're pushing each other to learn, grow, and be better. A handshake is thanking them for the opportunity to do what you love most. Of course, during the Covid-19 pandemic, the handshake is not recommended or allowed. However, players acknowledge each other by tapping their rackets in lieu of the handshake. The point is to acknowledge your opponent in recognition of a battle just fought. The same goes in business: to show respect for one another's intent to play fair and honest in the sandbox to get the deal done.

Sportsmanship is also a way of showing respect for the extended family and community of the tennis world—the fans. There are times when you are on the court that you can feel the passion and love emanating from the stands. Wendy Willis, one of my best friends since high school, would sit for hours in the searing heat of the first rounds of the Grand Slam tournament cheering on players of color. During the French Open in 2012, when Sloane Stephens made her big splash, Wendy was attending and took the time to sit in the stands to cheer her on even though she was in the middle of phone calls and paperwork for the adoption her daughter. In the early days, she went to all of Venus's and Serena's matches if she was at the event. Early on, Wendy sat through a five-hour match at Wimbledon to support Bahamian player Roger Smith. It was so hot out there, Roger was cramping from dehydration, "and I was sitting there, dying," Wendy told me. Everyone else left but not my

Wen. Later that evening, in the player's lounge, Roger walked up to her and thanked her for staying.

"You were the only one out there," he said. "Thank you so much for staying!"

My parents were so emotionally invested in my matches that it was hard for them to watch. My mother used to get so nervous when Dad turned on the sports news, she'd have to sit down and grip the edge of the kitchen table until the results were out. When I flew them out for matches, I had to beg Mom to sit in the front row. She'd tense up and cover her eyes until it was over or find a way to escape to a place where she couldn't hear the score over the speakers.

During one particularly brutal match at Wimbledon I was crying. Mom got up to leave because she couldn't handle seeing her baby girl upset on the court, or maybe I was just embarrassing her. I saw her and motioned to her to sit back down. I won the match, but my parents were wrecks. Afterward, Dad asked me why I was crying.

"Because I wasn't happy with the way I was playing and I wanted to win," I told him.

"You see what you put your mom and dad through!"

Shaking hands at the net is showing respect for all that our families, friends, coaches, and fans have given to us over the years through their love and support. It's acknowledging that when we play, it's not just about us. It's about something much bigger than ourselves.

I've found myself in multiple situations on the business side of sports where demonstrations of mutual respect have helped to diffuse the tension and find common ground on some important issues. But a sense of fair play was missing at a steamy and tense August 2017 annual general meeting of the International Tennis

Federation, the world government body of tennis, in Ho Chi Minh City, Vietnam, where hundreds of representatives of tennis from all over the world were in attendance.

When ITF president David Haggerty opened the meeting, he urged the delegates to work together for the betterment of the sport, including getting behind some major Davis and Fed Cup reforms designed to safeguard these flagship competitions. The idea was to combine events and shorten the Davis and Fed Cup season—all things the players had been asking for. There was a viable proposal on the table that would benefit not just the players but also the national tennis federations, which would get more sponsorship dollars and help with the development of the sport. Under the existing format, the growth of the event had been stagnant. Member countries weren't equally successful at bringing in revenues from advertisers, broadcasting, and ticket sales, with most running events at a deficit. But these national leaders weren't thinking bigger picture. Opponents to the changes also accused us of acting too hastily and trying to rush the changes through. They voted down the proposal.

The biggest underlying reason for the opposition to reform was that each country was attached to the "home and away" concept. They wanted more chances to host. The pride of being guaranteed to play in their own countries, in front of a home crowd, seemed more important to many than breaking even or making a profit. They were thinking about the short-term gains for their individual associations. It's human nature to dislike change after something has been done a certain way for decades, and there was fear that the individual countries wouldn't get as much income from the more compact format. They simply weren't ready for a different way of doing things. In many cases, it was the only time they could showcase professional tennis in their country, so it was understandable.

I got it. Being able to host your own Davis Cup or Fed Cup tie was essential for many players to have an opportunity to play in their home country in front of thousands of fans, players, and potential players, but the business plan wasn't viable for many.

Nevertheless, the benefits were obvious to me and far outweighed what would be lost. I was stunned when the reform was voted down at the Vietnam meeting. Everything came to a standstill. Voices raised and cheers ensued. I stared out from the dais at a scene of total confusion, then grabbed the mic that was in front of me.

"I'm extremely disappointed in your decision to not approve the reform and think differently," I said.

The hotel conference room went silent. For the next five minutes I had their attention.

"You elected us to be your voice and be accountable for our decisions. And yet when we bring you something worthy you don't trust it. You don't trust us. It's very disappointing to be sitting here as your representative right now. We need to think about what's best for tennis, our competition, our sponsors, and the needs of our players, not what's best for us as individuals."

That decision forced the ITF executives to go back to the drawing board, which ended up not being a bad thing after all, because by the next year we got a much bigger sponsorship commitment. By the next AGM meeting in 2018 in Orlando (the AGMs take place at different member-country locations around the world) we came up with a new and riskier approach, but by then we had more buy-in from everyone. After the previous year's fiasco in Saigon, the delegates knew they were obliged to keep a more open mind. They also had a full year to absorb the information and get feedback from players. It was an example well learned that in leadership things may not always go as planned, but if you can trust the process and be patient, there can be success.

The new format would involve a qualifying round in February, in which twenty-four teams take part in home-and-away matches—a key element of the Davis Cup's heritage. The twelve winners would secure a direct place into the final and join the four semifinalists from the previous year—who qualify without having to play in February—and two wild cards, which would be announced before the draw for the qualifying round.

So, the semifinalists of the Davis Cup would still be guaranteed a spot in the weeklong finals event in the following year and not have to play the first round in February. National associations would also benefit from lower hosting costs, including the reduction of match court availability by one day. That was a big deal, because the cost of hosting one of these tie events is steep. They must rent out the facility, then build it out to certain specifications. Costs are even higher in colder climates, where indoor courts can often only be rented by the week. They would need to pay the players, hire practice courts the day before the event, provide food for everyone, and hire extra service staff. If they didn't sell out every day, they'd lose money.

The proposed reforms would also help us reduce the number of dead rubbers—the matches in a series where the result has already been decided by earlier matches. If the fourth match is decisive and lasts three sets or at least ninety minutes, the fifth match would not be played; a match tiebreak will replace the third set in all Davis Cup dead rubbers, yet another benefit of the Orlando proposal. Above all, everyone stood to make four to ten times more money. The smallest country at the lowest tier went from receiving $2,500 to $25,000. Other nationals who were making profits of $100,000 went up to $400,000. That's what happens when you open your-self up to sponsorship opportunities. The Kosmos Group, led by soccer superstar Gerard Piqué, made the financial commitment to get the ball rolling, with brands like Rakuten, Lexus, and Louis Vuitton following suit.

Many players appreciated the reforms because they reduced the time commitment from four weeks, which is tough, to two weeks for those who had to qualify for the finals in a home-and-away format, or just one week if they were the semifinalists from the previous year. European and Australian delegates, especially, were initially against the reform, fearing fewer marquee tennis names would play in a weeklong event versus a weekend event, which could lead to waning interest from fans. But it turned out that in 2019, the first year the changes were implemented, the Davis Cup by Rakuten, which happened in Madrid, saw record attendance from both the top players and fans. Rafa was the star of the show, of course, and global tennis enthusiasts turned the Caja Mágica stadium into a weeklong party. Of course, year one had its chal-lenges, but we were off to a good start.

Not all the champions were immediately on board. But Rafa loved the idea from the beginning. "It is a good initiative that can work," he told the media. "It is obvious that when something does not work perfectly, you have to look for new solutions, and this has been going on for a number of years."

Novak Djokovic was even more enthusiastic.

"It is fantastic news. We all want to play for our country, but I have been saying for years that the current structure does not work. It will be more attractive for the world of sport, for sponsors, for media, and for fans . . . And there would be more money for the federations."

Everybody won.

Fair play, honor, and respect translate directly to many other arenas of life and business. How you conduct yourself toward others both on and off the court defines your brand. An integral part of this is cultivating an inclusive and open attitude. Being aware of diverse perspectives, respecting differences, and treating each other fairly in every situation elevates your game.

That can start with a handshake, which we use at the beginning and end of business meetings to show our intention to be good sportsmen and sportswomen throughout each stage of a transaction. Centuries ago, when people traveled with swords to protect themselves in strange lands, the handshake was used as a way of saying, "See, no weapon here, I come in peace." It's evolved to symbolize communication at its best, although we don't always live up to the promise of that initial gesture. Organizations often overlook the value of being transparent, inclusive, and responsive in their messaging. I can think of multiple occasions when communicating the right way, with the right tone and body language, has made all the difference in diffusing tension, ensuring team members feel respected and helping to get key stakeholders to embrace a goal or execute a strategy.

Good sportsmanship in business also encompasses principles that might not make sense if your sole focus is on quarterly profits. Avoiding negative sales pitches that attack the competition, staying away from the nasty rumor mill about other people in your industry, or not doing anything to undercut your rival or close a sale will do more to protect your long-term reputation than harm a short-term P&L statement. It will also build a loyal customer base over time.

Zappos, the digital shoe company, helps customers find products from their competitors if they don't have it in stock. Its CEO, Tony Hsieh, is known for saying, "Doing good is good business," and, as the erstwhile No. 1 seller of shoes online, he has a point. He has been focused on building a socially conscious business that is as responsible to its community and employees as it is to its shareholders. His emphasis is on maintaining a happy, healthy company culture, even if that means giving up the quick profits.

Another, earlier example of sportsmanship in business comes from the computer software industry's fiercest rivals—including Microsoft, Oracle, and IBM—who famously got together in the late '80s to combat intellectual property theft through the Business Software Alliance. They knew that banding together for mutual support instead of abandoning each other to the piracy wolves was for the greater good of the industry.

The most successful businesses understand that the rising tide lifts all boats, just as the champion on the court knows that feeling of doing the right thing far outweighs the pain of losing a point. In fact, corporate social responsibility is becoming a thing. More and more we see companies branding themselves in ways that have less to do with the products they are selling and more to do with giving back, like the yogurt company Chobani, whose founder and CEO, Hamdi Ulukaya, insists on hiring refugees to work in his business and is known for paying off school lunch debts for children in his community.

There is even a term for this sportsmanlike behavior in business: *conscious capitalism*. And it's being embraced by some of the most powerful investors in the world. BlackRock, Inc., a global American investment company that manages close to $7 trillion in assets, recently announced that it expects the businesses it invests in to have a positive impact socially and environmentally. In other words, they must build a sense of higher purpose into their business models.

"Purpose is not a mere tagline or marketing campaign," BlackRock CEO Larry Fink, wrote in a letter to CEOs in 2019. "It is a company's fundamental reason for being—what it does every day to create value for its stakeholders. Purpose is not the sole pursuit of profits but the animating force for achieving them."

As I contemplate my own transition from the USTA, continuing to be a sports analyst, expanding as a global speaker, and moving into a more corporate setting as a board member, that is the kind of culture I intend to be a part of; one that shares the values I have, that doesn't sacrifice what I morally believe in. Everything and anything that you do must have conscience. That's where I believe my experience on the tennis court and in global sports organizations has led me.

Of course, there isn't enough room to list the occasions in business, sports, or politics where that sense of altruism and sportsmanship *doesn't* exist. But I'm an eternal optimist. I believe the next generation of players will bring the spirit of true sportsmanship back into the arena, despite the lousy example being set by many adults. At the junior level we are seeing more players spontaneously shake each other's hands or show each other signs of physical affection during practice and matches. Introducing the pre- and postgame handshakes to our youngest players is building that sense of trust and mutual respect into their DNA.

Harlem Junior Tennis and Education Program graduate Justin Holmes has lived and breathed good sportsmanship ever since. But it was a life skill he had to learn. Justin was a lot like me when he was eleven years old—a perfectionist who took his losses hard. So hard, in fact, that he broke one too many rackets at a tennis camp, tapping his second-to-last one on a net post during a game changeover. Realizing he had only one racket left, that night Justin asked his father for more.

"You won't be getting a new racket until you fix your attitude," Holmes Senior told him.

On another occasion, Justin was doing drills on the Armory courts in Harlem and missed too many balls. In frustration, he hit a ball into the stands. His coach made Justin sit it out until he could regain his composure and respect for the game.

Today, Justin is so much more mature than his peers at Boston University, where he's studying medicine. He brings a level of courtesy and professionalism to all his interactions with others, to the point where people are astonished when they learn he's only twenty years old. It's a quality he attributes directly to his court training and character building in Harlem.

"Even if you're competing intensely with your opponent for hours, afterward you are both human beings and students of the game," he told me. "You walk in each other's shoes."

Now if only we could spread that message to our world leaders. As Donna Yuritic, the coach who noted my pouting problem when I was nine, put it: "We're working on it." But it took a female coach to have the confidence to tell me—gender equality is just as important as racial equality. Women have a lot more confidence speaking candidly to girls and women than men do and vice versa. It's only natural.

10

TWICE AS GOOD

To err is human.
To put the blame on someone else is doubles.

—UNKNOWN

In January 1989, Zina Garrison and I were playing in the doubles finals of the Toray Pan Pacific Open in Tokyo, but the singles finals ran long so we got off to a late start. This was a problem, because our flight back to Houston was due to take off at 7:30 that same evening. The drive from the stadium to Narita Airport could be anywhere from forty-five minutes to two hours depending on traffic, so the plan was to win, and win fast. But our match against Mary Joe Fernández and Claudia Kohde-Kilsch turned into a two-and-a-half-hour-long marathon, going into a third-set tiebreaker, which we finally won, 7-6.

Even though it took place in an indoor stadium with artificial grasslike courts, it was some of the finest tennis we'd ever played together, with moves that were perfectly synchronized as Zina set up shot after shot for me to put away at the net. We'd already had a great week in Japan. Tokyo was one of the cleanest cities that I had ever been to, and the locals were incredibly polite. Every time I got into a taxi I felt like a VIP because the drivers wore immaculate white gloves and had white covers on their seats. They were light years ahead of the rest of the world when it came to practicing cleanliness in public spaces. Their winter weather was cold and crisp, which, having come directly from Melbourne's 100-degree temperatures, felt refreshing and gave me extra energy. I played hard on and off the court by day, then explored the city, enjoying the Ginza nightlife singing karaoke, drinking Asahi beer, or at least I was, and eating sushi, which I'd first tried a year earlier and was only just beginning to appreciate. Zina was more reserved and didn't partake in a lot of sightseeing adventures. There's nothing like trying a cuisine in its home country. Despite running around during my off hours like a euphoric tourist, I returned to the stadium for training each morning feeling fresh and feisty.

Japan was fast becoming one of my favorite places. We were there just two months prior to play the Bridgestone Doubles Championships, hot off the Virginia Slims Championships in New York's Madison Square Garden. All the top female doubles players in the world had piled onto a Japan Airlines flight the following night.

By then, Zina and I were the No. 3 team in the world. But our opponents for the finals match, Robin White and Gigi Fernández, were formidable. The pair had won the US Open Women's Doubles that same year. Gigi, winner of seventeen Grand Slam tournament doubles titles and a two-time Olympic gold medal winner, would reach the world No. 1 ranking in Women's Doubles three years later. But not this time. We were on fire. It was probably the most energy we'd ever had. For one continuous week it felt like we couldn't miss. We won 7-5, 7-5. It was the end of my rookie year as a professional player and that was the biggest check I had ever received—half of $75,000. Yep, I was loving Japan.

But this time around there was no time for reveling, at least not for Zina. As soon as we heard "game, set, and match," she shook hands with our opponents and the chair umpire, then raced off the court so fast she almost gave herself carpet burn. Newly engaged, she was determined to make that flight home if it killed her, even if it meant missing the chance to hold up her trophy and join me on the podium for the presentation. She didn't even stop to take a shower!

"I can't miss this flight; I can't miss this flight. . . ." she kept whispering to me on changeovers.

Zina was so in love. She'd already been away from her man for weeks while on tour and she couldn't bear to be away from him one more night. They'd gotten engaged over Christmas and planned an engagement party for the day after she was back, and there was no way she was going to miss that moment with friends and family. And, oh yeah, a VIP appearance at a Prince concert. I would have rushed back for that too. Our coach had already gone ahead of us to check our bags and wait for us at the gate.

But there was no way I was going to miss *this moment* as the doubles champion, even if it meant enjoying another night of sushi and catching a later plane, the following day. I was going to grab that trophy and milk our win for all it was worth before I packed up my rackets. Plus, it was the respectable thing to do, as these fans supported us all week long. We couldn't both be MIA when it was time to say thank you. Standing there by my lonesome, I gave a speech on behalf of both of us, apologized for Zina's hasty departure, and made some wisecrack about how she had a hot date back in Texas. Then I trotted back to the ladies' locker room, showered, changed, and sauntered over to the car and escort vehicle, making it to the gate a mere two minutes before the plane door closed. When I took my seat next to Zina, we both burst out laughing. That was a trophy relay race for the history books.

A tennis doubles team, a marriage, a business partnership, or any successful relationship, requires an undying respect for the differences of the other person. But particularly so in tennis, where any fracture or weakness in the bond during those intense win-or-lose moments is going to be reflected in how you play the game. A male figure can interfere with the sisterly bond. It shows up clearly in the playing, with missed shots, grim faces, and plenty of glares. They say tennis is not a team sport, and it isn't, unless you're playing Davis or Fed Cup doubles or World Team Tennis. In singles, it's you and you alone on that court, with no one to high-five when you win a point, no one to coach you, support you, or encourage you while you're in the match. But in doubles, it's the two of you against the world, so you'd better be tight!

It's essential to be able to collaborate. You can't always insist on controlling the ball. Stepping aside and allowing my teammate to cover his or her area of the court requires a level of trust and humility. Because doubles is about working together and covering the entire court as needed and communicated between the two of you, tension will manifest in a lack of compromise as each player silently plays their own side. This is guaranteed to have a disastrous outcome, whether it's with the loss of the match or the loss of respect for each other. That's why it's important to maintain a good, trusting relationship on and off the court. Pay attention to how your partner is feeling and show up for them when you are needed. If my partner is serving for us to stay in a match, for example, and I know she is anxious, I'll plan to poach, intercepting the ball at the net and being more aggressive to take some of the pressure off her.

Being on a strong doubles team has taught me a lot about how to treat people in any situation. Great partners practice empathy. They are aware of how their actions affect others. One of the worst things you can do when playing doubles is look peeved or make an exasperated gesture. A curse word or an eye roll to let your partner know you're frustrated with how they missed a ball, for example, will do far worse to your game than losing a point. They know they messed up; there's no need to rub it in. I always tried to be positive and not show disappointment, because I didn't want my partner to treat me that way. Control your reactions, and whatever you do, never, ever gasp. It may be your turn to mess up next, so you brush it off, offer a wink of encouragement, become the stoic and steady rock, the jokester, the spirit booster with a fist bump, or whatever it is your partner needs you to be to fight their way through those inevitable slumps.

Of course, it doesn't always work that way. Of course, everyone has a bad day. There's no such thing as a perfect relationship. But there have been some disasters. It's not unheard of for doubles partners to go after the same guy. You'd be surprised what can happen to break up a team on tour. Just like you can tell when a marriage is in trouble, I've seen doubles partners squabble and hiss at each other courtside on plenty of occasions. It doesn't necessarily mean they'll lose, but the wins aren't consistent or sustainable. At the very least, these players need to set aside whatever is going on between them personally and get their game faces on so that their opponents can see them as a unified front. You've got to trust one another enough to know you'll get through it and fight through the storms.

It's not surprising that some of the all-time greatest doubles teams are siblings. No matter what, they're family. Venus and Serena have contrasting personalities and styles, but they know each other better than anyone, and they have a deep bond of history, love, and sisterhood that comes across in their doubles game. Identical twins Bob and Mike Bryan, known as "the Bryan Brothers," were practically unbeatable in Men's Doubles, having both held the world No. 1 doubles ranking for 438 weeks straight—longer than anyone else in history. Commentators attribute some of their success to the fact that they are "mirror twins"—Mike being right-handed and Bob being left-handed— enabling them in their coverage of the court. But I say their success comes from their positive personal connection. They say that identical twins have a kind of telepathy and that may or may not be true. But the positive energy between them is obvious when they celebrate each point with a midair chest bump! It probably doesn't hurt that they started playing doubles in the womb. The twins often say that they got an early start.

Martina Navratilova and Pam Shriver also exemplify the chemistry of a great doubles partnership. It's not always a given that a great singles player can do well as a doubles partner, yet these two champions came together to win a total of twenty Grand Slam tournaments, including four US Open, five Wimbledon, four French Open, and seven Australian Open titles. In the early '80s they had an unbroken string of 109 consecutive victories. A couple of their rare losses were against me and Zina, and once, when I played with Mercedes Paz from Argentina. In fact, I won with different partners against Pam when she wasn't playing with Martina. They were so much stronger together. I also won with Pam, the only event we played together, in 1989 in San Antonio at the US Hard Courts.

They had contrasting game styles. But their partnership worked because of the profound trust and confidence they had in each other's abilities. I like to think I rose to the occasion when I had the privilege of going up against the top-ranked champions. It was awe-inspiring to watch how perfectly synchronized they were.

Zina was the one the press anointed to be the next Althea Gibson—when she was the number one ranked junior in the world in 1981 after winning the Wimbledon and US Open Junior titles. Back then, African American tennis players were even rarer than they are today, so both Arthur Ashe and Althea made a point of recognizing and mentoring Zina. It was a heavy mantle to carry, not that she ever complained. At the time I had no idea how much pressure she was feeling to honor their legacy.

When Zina was fifteen, Althea told her, "You need to work on your serve."

At first, Zina didn't know what she meant. Then, "I realized she was actually talking about serving others and not just serving myself," Zina said. "And that I need to pass on the wisdom . . . especially to African American tennis players."

In fact, she did need to work on her actual serve. She had great placement but not a lot of power, so I can see how she might have been confused.

In December 1987, when word had gotten out that I was leaving Northwestern to join the WTA Tour and that I would be going to Australia, Zina's coach called to ask me if I would like to be her doubles partner.

"Oh my God, seriously?" I shrieked, jumping up and down like a little kid opening presents on Christmas morning the second I hung up the phone.

I was so excited that I told everyone. Then, two weeks before I was due to fly to Melbourne, the coach called.

"I'm so sorry, there's been a mistake. Zina had already committed to another player."

I managed to find a last-minute partner, Penny Barg (Mager), whom I'd played both with and against on the ITF Challenger Circuit. We did well, making it to the third round of the Australian Open. Penny and I also had success at other tournaments that year. Zina took note and this time made a firm commitment to play with me as soon as we got back to the US, at the Virginia Slims of Boca Raton, Florida. We went into it unseeded, playing some of the best players in the world, and we won.

There was a comfort level on the court from the beginning. It also helped that we were both black. It automatically gave us a sense of comfort in how we communicated or expressed our expectations of each other without adding pressure. We instantly felt a sense of sisterhood. We shared this never-say-never attitude, with a deep admiration for each other's game. Again, Zina was more the set-up person. She got the balls down low over the net and didn't miss, so I could be aggressive at the net and finish the point. Our slogan was Low/Go. If she hit the return low, forcing our opponent to have to hit the volley up, I would go, cross the middle for the poach, and hopefully put away the shot. I was fast where she was nimble. If I happened to be out of position or overplayed something, Zina was quick enough to run down certain shots and keep us in the point. I was the eager rookie while she was the older sister and mentor.

I was so green that during our first match together I almost started to hyperventilate on the court. We were down a set and I had to pee so badly I felt like I was going to burst. The facilities were far away from the court, and I didn't know the rules. I had no idea I could leave and go to the ladies' room during a break. Seeing the look of panic on my face, she asked, "What's wrong?" When I explained, she laughed and said, "Well, go!"

I came back so calm and focused that we won the match in three sets.

Zina and I had each other's backs in every way, on and off the court. You experience so much on tour with your partner that you develop a book's worth of inside jokes. Because we were both black, we understood each other on a whole other level. Our lingo was different, and we communicated with ease. We shared a deep sense of passion and understanding, which no doubt allowed us to be ourselves at all times. We could be unapologetically direct with each other and not have any hard feelings, because it came from a place of love and support. We knew we represented something greater than ourselves.

Zina was always nervous whenever we played in her hometown because she knew all eyes would be on her. It made her a little too conservative in her playing, which put us at a disadvantage. During the Virginia Slims of Houston, our coach Willis Thomas was trying hard to get her to poach and intercept a ball. It would have been an unusually aggressive play for her whenever she was nervous, so I joked that if she poached at a critical junction in the match, Willis would fall off his chair. She did, and so did he, completely shocked that she made the move in that moment. We couldn't stop laughing.

When you work and play with someone 24/7, hopping planes and sharing hotel rooms in every tour location, you can find yourself in some absurd situations that no one else would believe if they weren't there to witness them—like when Bootsy Collins started stalking Zina at the 1988 US Open.

James Brown's former bass player and one of the leading members of the legendary late '70s funk band Parliament-Funkadelic, Bootsy wasn't exactly Zina's type. My girl is conservative and shy, almost to the point of being an introvert. Fame was thrust upon her because of her incredible talent, but she never sought the spotlight. If anything, she ran from it. Bootsy, on the other hand, was outrageous, with a collection of alter egos from Casper the Funky Ghost to Bootzilla—bizarre alien rock star bedazzled head to toe in rhinestones. You may remember his onstage presence as the man with the big hair, a glitter top hat, and starburst sunglasses.

But their differences in style didn't seem to matter to the funkmaster. Bootsy was coming out of a limo with Miles Davis on Central Park South when Zina caught his eye. She was sitting at a restaurant around the corner from the Essex Hotel in Manhattan where she was staying, while he made some calls and tracked down her hotel room number. He kept asking her out, even offering to buy her a tennis bracelet. The whole thing was like something out of that Looney Tunes cartoon, Pepé Le Pew. Zina was the desperate little cat who was always trying to get away from the amorous skunk.

Though categorically not interested, she was kind enough to arrange for him to get courtside tickets to our doubles match, never actually expecting the man to show up. Not only did he come, he dressed in his full regalia and brought along his posse. He arrived a little after the match had started, but you couldn't miss him in the grandstand, especially because he was making such a fuss over his seats, which he didn't feel were close enough. Zina was mortified.

"He's heeeere!" she informed me on a break between points, using a weird combination of a whisper and a shriek.

"Well, you can't miss him!" I said, then we both burst out laughing.

Bootsy, more than game strategy, became the subject of our many conversations between points and sets. We were both hysterical, especially when we recognized Bootsy's voice cheering above the crowd every time Zina won a point.

As soon as we won the match, Zina went tearing back to the women's locker room. But this was long before all the renovations at the Flushing Meadows facility, so she could only get there by running through the old Louis Armstrong Stadium, when the Grandstand Court was connected to it, possibly risking bumping straight into Bootsy. It was my job to run interference and distract him so that she could make her getaway.

Being a fan of the Funkadelics, I didn't mind one bit. Zina knew that walking straight up to famous people and introducing myself was kind of my thing. Bootsy and I had a nice conversation while Zina made her getaway. I'm happy to report he smells nothing like Pepé Le Pew.

My time as Zina's doubles partner came to an end in 1991. I wasn't ready for us to go our separate ways because I believed we had more wins in us as a team, particularly a Grand Slam title, and was extremely disappointed when she told me that she was playing with someone else the following season. Even though I was pissed, I got over it because at the end of the day it was business and had nothing to do with our friendship. Zina knew then what I know now: not every partnership is supposed to last forever. She'd taught me all she could in my first years of going pro, and I had taught her a thing or two. My next stage of growth had to come from elsewhere. I had to accept that even the best of relationships has seasons and move on to the next point. We partnered again in 1995, her last full season on the tour.

I then partnered with Lori McNeil, whom I played and trained with on occasion when Zina and I weren't at the same event. We had a great winning percentage together and enjoyed each other's company. Lori is a complete jokester and always had me laughing, which was healthy for the partnership. But for some reason, once we committed for the yearly partnership, we couldn't win the titles. We reached several finals but only got so far, so we split the following season.

By then I'd already developed friendships with many other champions and had established myself as a top player. You get to size each other up after years on the tour, especially when you play regularly against each other. I always made a point of treating everyone with camaraderie, courtesy, and respect, because you never knew when your competition could become your greatest ally. (It's the same in business. You never know who is going to end up being your boss or colleague.)

Manon Bollegraf and I hit it off from the moment we met, which certainly helps when you're a WTA doubles team. We'd played against each other a couple of times, so we knew each other's style and temperament. We also started sharing the same coach, Charlton Eagle, a South African, Australian person of color. Neither of us remembers who asked who to be partner, but as with Zina, we made the perfect complement. She set me up so that I could go in for the kill at the net, unselfishly allowing me to shine. Bollie was the reliable and steady one who enjoyed what she described as my "flashy and erratic" approach. The yin to my yang, she relished the role of being my enabler on the court and celebrated our differences.

"You are my black sister," she once told me. "On and off the court."

"Well, you are my orange sister," I told her, orange being the national color of Holland and my favorite hue next to purple. It's so important to appreciate someone's culture. Just because you are from different parts of the world or have different skin tones doesn't mean that you aren't similar. Appreciate the differences.

When I was on tour the Netherlands felt like my second home. Often if we had a long European tour, I would go back with Manon to her home for a week off and train there. It was a lot easier and cheaper than flying back to the US just for a week or two. There's a kind of authenticity to the people and the place feels familiar. I've always loved the direct, no-nonsense approach of the Dutch, and Bollie didn't disappoint. She was a straight shooter like me: honest almost to the point of bluntness and never catty. I liked that, because it meant we never had to worry about mind games. Clear and open communication is another ingredient for a successful partnership. You must communicate to dominate on the court, no second-guessing each other. We had that complete transparency with each other from the beginning.

More important, we also had fun together, whether that was catching musicals in London or eating our way through all the Grand Slam tournament capitals, particularly Paris. Bollie shared my sense of adventure and was always up for some sightseeing or trying out new cuisines, particularly Asian food with a touch of spice—the hotter the better. She also allowed me to take her on some style adventures. We shared the same clothing sponsor, Nike, so I took it upon myself to select our outfits and match us right down to our socks, wristbands, and underpants. Bollie was never much for fashion, but she enjoyed looking good on the court.

One of my favorite memories with Bollie was playing a tournament in Birmingham—one of the lead-up tournaments to Wimbledon. It just so happened that the World Cup was taking place at a nearby stadium, where Holland was playing against Switzerland. Bollie got us tickets, taught me a couple of songs in Dutch, as well as a few curse words, then we went shopping for head-to-toe Dutch outfits. We had a blast cheering the men's team for Holland. When we were in Houston, I returned the favor and introduced her to basketball. We went to some Rockets games and met a couple of the players who happened to be friends of mine.

As much as I admired and appreciated Bollie's cool and calm

temperament on the court, I never understood how she managed to avoid being ruffled during a match, no matter what was happening around her. Then one night during the US Open, while we were sharing a room at our agent's house, in twin beds that were only a couple of feet apart, I got my answer. Bollie started talking in her sleep. Well, it was more like sleep-shouting, and I even managed to make out one of the Dutch swear words she'd taught me. Bollie was so worked up, her legs were moving under the sheets as if she were running back to the baseline to return a ball. Then she stood up on her bed and started sleep-running over to mine and kept going. I didn't dare try to wake her up, as I'd heard that you're not supposed to wake up sleepwalkers. But I was amazed she didn't crash right into the opposite wall! Instead, she stopped in stride.

The next morning, we laughed so hard our sides ached.

"Now I know the secret to your self-control!" I told her. "It all comes out in your sleep!"

In 1996, the time came for Bollie to move on. She was an uncharacteristic bundle of nerves when she blurted out the truth on the side of an indoor practice court:

"Kat, we need to talk. Martina has asked me to become her doubles partner. But I've already committed to play with you next year, so of course I'll tell her no."

"You will do no such thing," I told her. "Bollie, you've earned this. What a fantastic opportunity!"

"You don't mind?"

"If Martina asked me to play with her, I would be the one coming to you with this conversation. You *have* to go for it!"

It was an incredible honor and gift for Bollie to be invited to join as that icon's partner. By then I'd learned that even the closest of partners don't always stay on the same path. Great leaders set people free to grow, learn, and flourish, even when that means a moment of loss or inconvenience for themselves. We'd had a good run, but we weren't quite getting to the finish line at the Grand Slam tournaments. Maybe our progression had leveled off and this change would be good for both of us.

It thrilled me to see how much Bollie benefited from the experience of playing with Martina, having reached the No. 4 ranking as a doubles player. Obviously, she wasn't playing with her the whole time, but the impact was long lasting. By the time Bollie retired in 2000, she'd won eleven Grand Slam tournament doubles titles, including Mixed Doubles.

"Playing with Martina taught me so much about myself," Bollie shared with me years later. "I felt a huge amount of pressure on me, because I knew that if we won, it would be everyone's perception that it was because of Martina, and if we lost, it would be because of me. Martina picked me because she wanted me to call the shots in certain situations. She wanted me to dare to be the leader in our team. Like a great champion, she knew what her strengths and weaknesses were and trusted me enough to step in when she showed her vulnerable side. It was only one year, but it was extremely important for the rest of my career."

I've learned a lot about leadership and partnering from all my experiences in playing doubles. Although I've always been independent, life is a little sweeter and easier when you can share it with someone.

The most important partnership of my career outside of tennis has been with my best friend Roberta. Beyond her business insights, Roberta has balanced out my hard-driving ways with a sense of adventure. It would never have occurred to me, for example, to travel to Cambodia or pop over to Iceland for a soak in the hot springs if she hadn't given me a nudge. Because she is so inquisitive, Roberta does the research, figuring out which excursions she can tack on at the end of our global meetings enabling us to not only recover from an intense workweek but to view other cultures, which I would then relate to how I view the world, to better understand my craft and goals. Thanks to her, I've found myself riding elephants, petting tigers, and trekking through the foothills of the Himalayas. If it wasn't for her, I'd have been all business on these trips and missed these opportunities to explore.

After speaking with Kate Middleton, the Duchess of Cambridge, about Bhutan and learning from her that it was considered "the happiest place on earth," Roberta came up with the idea of traveling to Bhutan. We went with our usual companions on these excursions, our golfing buddies Charlotte Kuey and Lisa Grain. Charlotte calls us the "Golfernistas." A few days into our trip, our tour guide told us a fable about four harmonious friends: an elephant, a monkey, a rabbit, and a bird. The story goes that the bird picked up a seed and planted it. The rabbit watered the plant, the monkey added fertilizer, and the elephant guarded it until it could grow into a tall, majestic fruit tree. Interpretations vary, but the elephant represents the body, or strength; the monkey represents the mind, or wisdom; the rabbit represents emotions and speed; and the bird represents soul, or vision.

"You know, you ladies remind me of those four friends," our guide, Yeshey Tshering, told us.

We started debating about which animals we were. We decided that I was the elephant, Charlotte was the monkey, Roberta was the hare, and Lisa was the bird. Then I thought about the dynamic I had with my three other besties from college—Wendy Willis, Lori Shaw, and Lauren Lowery. We called ourselves the "Four Musketeers" because we've had each other's backs for decades.

But I couldn't quite pinpoint who was which animal because it depended on the situation. My relationship with all six of these women is the definition of interdependence. We leverage our differences into strengths, each one stepping up for the other for the benefit of the whole group.

During the rest of the tour we noticed little shrines to the four friends wherever we went. The most popular image was of the animals standing on top of each other, the elephant on the bottom, then the monkey, the hare, and the bird, which was picking the fruit of the tree and passing it down to its friends.

That was us.

11

REACH BACK, PULL FORWARD

*No matter what accomplishments you
make, somebody helped you.*

ALTHEA GIBSON

Even though our travel and meeting schedule can get crazy, when overseas I've always tried to see and experience something new along the way. I'd never met any Bhutanese players and knew nothing about the place other than the fact that it's a small Buddhist kingdom on the eastern edge of the Himalayas. The physical landscape of this landlocked terrain—which is comprised of lush green valleys rimmed by majestic, snowcapped peaks—took my breath away. I gazed up in awe at the tiered roof temples perched precariously on the mountainsides.

We were driving down one of the streets of the capital, Thimphu, when I noticed a sign: BHUTAN TENNIS ASSOCIATION.

Any tennis court makes a tennis player feel at home, so I asked the driver to stop so that we could go inside and check out the place. I also wanted to introduce myself. The way we were treated, you'd have thought we were royalty. They hustled about getting us tea, coffee, and snacks, calling their top leadership to tell them about our arrival. We'd only planned to stay about fifteen minutes, but an hour and a half later we were sitting down with their top officials, learning about some of the main challenges they were facing in their efforts to develop young Bhutanese players and being asked, as head of the USTA and VP of the ITF, if there was anything I could do to help.

It was taking them a while to get to the point, but after some persistent probing I learned that their main need could not have been more basic: some of the players were using rackets that had become worn down from overuse, and they didn't have a stringing machine. They needed to be able to maintain what limited resources and equipment they had.

"How much to buy you a new stringing machine?" I asked the officials. The only racket stringing machine in the entire town was so antiquated it looked like it should have been on display at the Smithsonian.

"Four hundred dollars" they finally told me, after some embarrassed throat clearing and shuffling.

"Done!" I told them.

I immediately organized to have a stringing machine shipped to them from Thailand.

Reaching back to pull people forward is why we're here. It's why as leaders we do what we do. It's also how I was raised. I noticed how my parents gave of their time and support to less fortunate children in the community, whether as educators, helping out the kids in their schools with pencils, notebooks, and other supplies, volunteering at the Dr. Martin Luther King Jr. Boys Club, or having half the neighborhood over to our house for a good meal.

Now it's my honor and obligation to do the same—providing guidance and giving back—and I wouldn't have it any other way. Everything I am was seeded in the love and commitment of my parents and people in my local community who saw something in me. They gave of their time, experience, money, and knowledge to nurture a gift that might have died on the vine had it not been for their collective attention and care.

Countless acts of kindness enabled me to pursue my dream. Kids in communities like the one I came from don't generally have the resources or facilities to practice and develop their game. The smallest things, like getting safely to and from a practice court when both parents have to work or paying for the right shoes, rackets, or entrance fees to play a tournament, may be enough of an obstacle to prevent someone with raw talent from ever getting to the next level.

But inner-city junior tennis in my hometown was very much a community enterprise. Through the generations, there's been a burning desire for inclusion in all forms of athletics, especially for those who don't necessarily have the means, whatever their race. Maybe it dates back from the "Negro" leagues—professional baseball leagues comprising mostly African American and some Latin American players—which came into existence when people of color were excluded from organized sports. In tennis, we have the American Tennis Association (ATA), the oldest African American sports organization in the US, which although inclusive of all groups today was formed because blacks were not allowed to compete in the Unites States Lawn Tennis Association (USLTA), which is now the USTA. This is where the likes of Althea Gibson and Arthur Ashe got their start competing in tournaments.

This desire to reach back and support young talent didn't necessarily have to occur under the aegis of the ATA. There was just this natural impulse to lift me up. From the guys who strung the rackets in the pro shops to the parents of the other kids in my school who weren't even necessarily tennis players, I had dozens of people in my "court." They were so hell-bent on seeing me flourish—they reached into their own pockets and organized fundraisers to ask others to do the same.

Julia Steele was prominent among those who got behind me and my family to make sure my tennis needs would always be taken care of. A math teacher in the public school system on Chicago's South Side, Julia fell in love with the game late in life, becoming a director of junior player development at the Chicago Prairie Tennis Club (CPTC). She often played at Lake Meadows Tennis Club, where I caught her eye as I was dragging a ball hopper to the last court to practice my serve. I was eight years old and the basket was almost as tall as I was.

"I've never seen a child your age so determined to practice," Julia told me.

Julia, or "Miss Steele" as we respectfully addressed her, along with other members, Joyce Clark, Ruth Leake, Richard Bradley, Beverly Coleman, and countless others, immediately started collecting money from the other members through fundraisers or direct contributions to the pot so that I could play in tournaments all around the Chicago area. Her motto was LOVE TO SERVE, which coincidentally is also the name of a prominent National Junior Tennis and Learning network chapter on Chicago's South Side, founded by Lamont Bryant and led by Lori James. She's dedicated her life to helping kids in underserved communities gain access to the sport. At the time, we juniors weren't always aware of all Miss Steele did for inner-city tennis. But we appreciated the spread of fresh fruit she always laid out at our events. (Actually, it was the gumballs she handed out to her favorite players that we loved her for most.)

Dr. Kim Williams was another member of my circle of support and an early instructor at the Boys Club summer program when I was first recognized as a talent.

"I'll never forget how you were rallying back and forth within thirty minutes of picking up your first racket," he told me, many years later.

I was fortunate enough to meet Kim during his last summer before starting medical school. Like me, Kim was an inner-city kid with little in the way of access when he first came across tennis. He had originally wanted to play baseball, but his parents didn't have the money for a bat, glove, or cleats. His aunt, who had played tennis back in the '50s, gave him one of her old rackets, which he used to practice against a concrete wall until he could volley and serve just about well enough to play for his high school.

Kim was determined to build his skills in tennis to the point where he could get an athletic scholarship to the University of Chicago and started practicing six hours a day. He got into his dream school, on a full-ride scholarship, making team captain his freshman year. But it had always been Kim's goal to become a doctor. Having experienced the lack of access to quality medical care growing up in the inner city, he wanted to make a difference. He started teaching tennis to put himself through medical school, and today he is chief of the Division of Cardiology at Rush University Medical Center.

It took someone like Kim to see something in someone like me. And it takes someone like me to see something in someone like Justin Holmes.

Justin was one of the kids I connected with in my role as

executive director of the Harlem Junior Tennis and Education Program (HJTEP). When I first joined in 2005 I was two years into my career at the Tennis Channel and based out of Bradenton, Florida. I was still coaching part time, but I'd been getting more deeply involved in the global and national tennis organizations, moving beyond the role of player ambassador to board member. I was getting an inside look into the sport at all levels, which made me realize that the need to provide more access to more diverse generations of players was not being adequately addressed.

The HJTEP had been around since 1972, when Claude Cargill, known as the "Angel of Harlem" and one of the first African American policemen in New York, along with fellow police officer Bill Brown started the program at the 369th Regiment Armory in Harlem. They knew only too well that few kids in the Harlem community had exposure to the mostly white-dominated and white-managed sport. They saw tennis as a means for betterment and a powerful character builder. They also took inspiration from the fact that African American tennis greats Arthur Ashe and Althea Gibson played at the Armory and would make great role models for young players.

Mr. Cargill and Mr. Brown had already been privately buying equipment and paying tournament fees for young players. Players themselves, they gave lessons, but when promising players came along that needed a higher level of instruction, they helped them by securing private coaches and, when needed, room and board. But the program really took off when former Knicks player Earl "the Pearl" Monroe got involved with fundraising. He, along with Bill Holloway, another tennis instructor, teamed up with Mutual of New York Financial Services to sponsor an annual invitational tournament to benefit the program. Dante Brown, an alum of the program, played at Southern Illinois University on a tennis scholarship, alongside one of the best American Grand Slam doubles teams of Ken Flach and Robert Seguso. He later came back to the program to be a coach, board member, and administrator for many years, to share the benefits of what being in the program could provide. The other coaches were also alums of the program, which was a testament to what this organization meant to them in terms of giving back.

I was drawn to the organization because it was that same community effort that raised me as a young tennis player in Chicago, but on a much bigger and broader scale where high-profile donors—politicians, athletes, and celebrities—are involved. Former New York City mayor and my second dad, David Dinkins, is a huge supporter of HJTEP, as he is of all things to do with tennis in the inner cities. Over the years, HJTEP has transformed thousands of young lives.

The founders' ambitions for their young players went much further than the sport. In 1979, they launched a Homework Club to provide one-on-one tutoring and academic counseling. When I joined, it was just the Harlem Junior Tennis Program, with no mention of education, although that was always a big part of what we did. We immediately changed our name to more accurately reflect all that we offer. Today the program provides computers and printers to assist with homework, and online tutoring modules. Led by our director of education, Ivy Leaguers, and program graduates, HJTEP helps high school students prepare for their SAT and ACT exams, as well as helping them with their college applications. We also offer guidance on financial aid and scholarships. For many students, the academic support we offer leads to higher education and professional lives. More than 90 percent graduate from high school, and more than 60 percent of our program graduates go on to college.

Beyond supporting our juniors in their pursuit of academic excellence, we also focus on emotional and physical wellness, because it's not just about developing great players. We want our kids to become well-rounded human beings with no limits to what they can achieve. Every child needs to understand that no matter where you start, you can rise to the highest levels, just as I rose from the public courts to the boardroom. But our program relies on funding from donors, fundraisers, and corporations to foot the cost of operations and scholarships for those in need.

Justin exemplifies how the support of a program like this can put a disadvantaged kid on the path to greatness. The only child of a single mom, Harlem Junior Tennis was his source of stability. Throughout his childhood and teen years, Justin and his mother had to move multiple times, often couch surfing with friends and relatives. But HJTEP was always his safe haven. We were like the extended family of brothers and sisters he didn't have at home. Justin even called me his second mom.

I quickly realized that my new role was as much about nurturing the whole person as it was about raising funds and running operations. Beyond coaching them to become stronger players, I spent a lot of time just listening as they talked about their hopes and dreams; the turmoil in their lives; and the challenges they were facing with school, their relationships, and their physical and emotional health. I later discovered that one of "my kids" was living in a car. I had no idea this was happening.

But they were okay as long as they had us as the second home they could come to—a place of their own where they could be accepted without judgment. My door was always open, and plenty of tears were shed in my office. Sometimes I had to display some tough love, but mostly I doled out hugs and assurances that it would all be okay. These kids opened up to me in ways they weren't always able to with their parents. I even knew who was crushing on who and found myself telling them to keep it as an innocent relationship with no hanky-panky allowed.

Justin's struggles were a little more serious than which girl to ask to prom. In his sophomore year of high school he learned his mother had cancer, just as he was going through the intense process of college applications, sitting for his SATs, and wondering how he was going to pay for it all if he didn't land a full scholarship. Bright, ambitious kids tend to put a heap of pressure on themselves as they go through their sophomore and senior years. They feel like their whole futures are at stake.

Justin had always been engaged and enthusiastic, eager to pass on what he'd learned to the younger kids in the program he helped to coach. From the age of fourteen he'd been a ball boy at the US Open. One of the few kids of color on the courts, he did us proud as the perfect ambassador for HJTEP. He met all the biggest tennis stars, and they were charmed by this poised young man. He did commercials for ESPN and interviews on the Tennis Channel. Justin had traveled all over the country for tournaments and tennis clinics, and he blossomed from the experience and exposure to people from different parts of the country and the world. I had a feeling about Justin and knew he'd go places. But something in his demeanor wasn't right. He'd been walking around with his chin in his chest, barely saying a word to anyone.

"Hey J," I called out as I saw him pass by my office door. "Come talk to me!"

"Hey Kat," he murmured, as he slumped into a chair in front of my desk.

I looked straight into his eyes and I asked how he was doing, and it all poured out. Justin shared about his mom's illness, his fears for the future, and all the crap he was taking from the other kids in his high school who were mostly jealous of the path to success he was on. Justin, whose father was a good tennis dad but did not live with them, felt completely alone. He didn't want to burden his mom with it all as she went through chemo and radiation, so he bottled it up, and now he was heading to a dark place of depression that could derail everything he'd worked hard to achieve. I feared his spirit was breaking at a pivotal time in his academic career. Knowing he needed more than my soft shoulder to cry on, I called a psychiatrist friend to speak with him and give him some coping tools to help him pull through to the other side. The experience also reminded him that his tennis family would always be there for him. With our unconditional support, he no longer felt so isolated.

Years later Justin is currently going into his senior year at Boston College, where he's studying biology and premed on a partial academic scholarship. He could have easily made the college team, and sometimes joins in as a practice partner, but he'd made a choice to focus on his intensive coursework. Like my old friend Kim, Justin's experience going through the health care system with his mom, who thankfully is now cancer-free, made him want to "do something to help the people of this world and make my own life better." Ultimately, he wants to become an orthopedic surgeon, inspired by the sports doctors fixing up injured athletes.

Justin credits the life skills he's learned through tennis for the successful transition he's made to college out of state. It's instilled in him a sense of autonomy, self-discipline, and a desire to follow new opportunities wherever they may lead him.

"Making your own decisions and being accountable to yourself are things you learn so well in tennis," Justin texted me recently. "Tennis is not just a sport and HJTEP is not just a program, it's a way of life, and I can't say where I'd be without the coaches and players I've grown up with." "HJTEP is not just a program, it is a way of life," was our original slogan. This obviously resonated with him.

That's exactly what Steven Wilson would like to be able to go back and tell the kids in his neighborhood in the Bronx who used to tease him for walking around with a tennis racket in his hand. One of seven children and the youngest of five brothers, he was just five when he and his father took the No. 2 train down to the HJTEP. From the moment his racket strings first connected with that ball, "I fell in love, hard," he told me. That's probably why I always felt such a strong connection to him.

When Steven wasn't learning the game on our courts, he was practicing every spare hour of the day on a local handball court. But he took a lot of flak for his passion.

"Tennis is not a black sport," the other boys used to tease him. "You should be playing football, baseball, or basketball; tennis is so dumb."

Steven also competed in track and field, but as the years went on it became increasingly clear that he had serious star potential in tennis, and we encouraged him to focus and see how far it could take him. One of the best competitive players from the HJTEP, Steven used his athletic discipline to earn a tennis scholarship to Wilkes University. I was so proud of the kid I took him on a college shopping spree. He couldn't afford to buy all the proper prep items for college, so our chairman, Jim Kelly, personally footed the bill. Jim, as I do, strongly believes that no child entering college should be without the essentials for survival and to fit in. We had fun buying linen, dorm lamps and accessories, preppy clothes, a laptop—everything Steven might need for his first year at college and away from home. Another reason why donors are needed to fulfill these needs.

Steven majored in business management and minored in marketing. He had played well for Wilkes, making the nationals all four years of his college career. But as his college career went on, he was beginning to see his future in business rather than professional tennis.

"You've given me so much that I want to get the most out of being here," Steven told me. "I want to win for you academically."

At just twenty-two, Steven became New York division manager of a private tennis program based in Tribeca that he hopes one day to take national, then global. In this role, he gets the best of both worlds, applying all the skills he's learned from tennis to his business while still being physically around the sport all day.

"I was able to grab examples of leadership from everyone, learn how to go out into the world, manage people, all by watching you guys closely and taking notes," Steven told me recently. "You showed me how to conduct myself on and off the court."

Coaching was another one of his gifts, so we put him to work in the program from the age of seventeen, and he came back every summer through college to help us out. As great an athlete as he was, Steven shone as a mentor to the other kids. We took him on coaching expeditions to Florida, demonstrating how, as a leader, he could adapt to each situation: donning his game face when it was time to oversee a practice or tournament; kicking back and relaxing with the kids when it was beach time; or engaging and networking at dinner with other coaches, sports executives, and sponsors of tennis. Looking back on Steven's earlier struggles with self-esteem, it was beyond gratifying to see how much he'd come into his own as a confident young man.

I recently asked him how the sport has changed his life.

"Tennis literally gave me the keys to everything I am doing now in life. It's not just the sneakers, socks, shorts, rackets, strings, grips, tournament fees, airfares, and hotels that HJTEP gave me. It's the friends I have, my college career, my job, what I do in my off hours. It all goes back to Harlem. I don't take it for granted; I am forever in its debt."

To a person, my HJTEP graduates tell me that, beyond the financial support and the coaching they received from our staff, the game's greatest gift to them has been the life skills it's taught them, even when their life's path has taken them in a different direction from the sport. Vashni Balleste, who is now chasing her dream of becoming a film director in Hollywood, put it this way:

> Being on the court, you don't know what's going to happen, so you learn to come in with confidence and a certain mental toughness. It doesn't matter how much more skilled you are than the other person; you can tank a match with your mind.

HJTEP taught Vashni the importance of having a personal routine of necessary steps and practice to reach her goals. Vashni, Steven, and Justin, like the tens of thousands of others who have benefited from the program, learned that they can do anything with a combination of structure, work ethic, and courage to test the limits and see how fast and how hard they can hit it.

When Justin graduated from high school, he was given the Fitzpatrick Trophy for public service. The recognition came with a cash prize of $1,000, money that could have gone toward school supplies and all the other expenses of starting in your first year of college. But instead of pocketing that money, Justin decided to donate it to HJTEP. I asked him why.

"When you are disadvantaged, sometimes a little help or opportunity is all you need to actualize your potential and make it into something real," he explained. "Having been given that help, the cycle has to continue."

Donating that money made Justin feel empowered, as though he was walking in the footsteps of the people he looked up to. Justin's goal was to make sure the next generation of kids coming out of HJTEP will be superstars and go even farther in the game and life than he has.

Giving back is addictive. The more you experience that gratification, the more you want it. I'm not referring to corporate checkbook philanthropy; that's something you do because you're supposed to. I'm talking about getting out there, being hands on, raising funds and awareness for the communities and causes you are most passionate about.

My first taste of actively giving back to someone happened in 1995, at the end of that year's Virginia Slims Championships. Lee Jackson, one of the den mothers on the tour, had always expressed a desire to go to the Australian Open. She'd done so much for us over the years, I decided it was time she went, so I invited the twelve women in the tournament—some of the most successful players on the circuit who weren't lacking in funds—to contribute $1,000 each, enough to cover Lee's business-class ticket to Melbourne and a nice hotel room, with some spending money left over. Of course, everyone was happy to chip in. When our Madison Square Garden tournament ended, we all headed back to the UN Plaza Hotel where we were staying. We'd organized a reception in someone's room, where we presented Lee with a letter, telling her how much she was appreciated, a card signed by all of us, and the travel agent certificate. After seeing the tears of joy on Lee's face, I was hooked.

Early on in my tennis career, Zina Garrison instilled in me that helping the generations that come behind us is not just an obligation; it's fun. I'd even go as far as to say that the joy I've gotten out of giving almost feels like a selfish pleasure.

Zina had learned well the importance of reaching back from those who came before her. She made a point of supporting younger, gifted players of color, which was why she invited me down to Houston to play with her and Lori McNeil during the winter of my sophomore year in college. Talk about a confidence builder! Later, when we became doubles partners, we used to throw fundraisers for inner-city tennis for the Zina Garrison Foundation at her home in Houston with a guest list that read like a who's who of sports and popular culture. It was a chance to dress up and be among people who were just as passionate about creating opportunities for young players as we were.

Particularly as an African American who has been the first and only in many situations, I feel it's my duty to reach back and bring the next one forward. I was once that inner city child with a dream, having no idea how to achieve it other than through working hard and enjoying the process. It took a village to raise me, support me, uplift me, and allow me to shine. It wasn't perfect by any means, and I surely didn't know what my path would ultimately be, but I learned that I was never alone and always appreciated that fact.

That's why my mission is to give back wherever and whenever I can to make a difference and perhaps inspire a positive trajectory in a young person's life. HJTEP gives me the opportunity to help the next generation become leaders of their own destiny. As people of color we are extremely aware of our good fortune, and we always try to do our part to support each other and pass it on.

Here's another selfish reason to give back: it's good for your career. More and more, businesses are integrating philanthropy into their overall business model. Doing something for the larger good is good for the brand, makes shareholders happy, and boosts the morale of the workforce, who feel a greater sense of purpose. As I speak with more executives and take my seat on more corporate boards, I notice more conversations about this.

But giving doesn't necessarily need to be on some vast corporate scale. Whether we know it or not, we all have experiences and of-ferings that can benefit others. In giving that wisdom or empower-ing others with resources that can transform their lives, we enrich ourselves. I have a wealth of knowledge as a player and a coach that I need and desire to share with others, especially kids in need. Peo-ple showed they had an interest in me and believed in me, which in turn helped me to believe in myself. It's why reaching back and pulling forward has become my whole purpose in life.

There are no limits to how you can give. In reaching back, you can also stretch your hand halfway across the world.

After an annual general meeting for the ITF in Vietnam in 2018 I decided to do an excursion trip to Cambodia. This time I'd already met the Tennis Cambodia director at the conference, who formally invited me to their facility in Siem Reap to see their junior program. One afternoon, in between touring the temples of Angkor and riding down the streets in a *tuk-tuk*, I went to the address the director gave me to find two tennis courts at a hotel resort where the program was headquartered. I saw about a dozen children of all ages and levels playing with passion, laughing, and cheering whenever one of them hit a good shot.

I couldn't help but notice that their tennis clothes were not ex-actly appropriate. Some wore flip-flops. No one seemed to own a pair of tennis shorts. Many of the kids only owned one outfit be-yond their school uniforms.

I did some drills with the children and young adults, speaking with them individually and encouraging them to stay in the sport and use the life skills they were learning to advance in their education and careers. In addition to the life advice, I wished I had a bag of tennis gear to give. All I had was a pair of sneakers, a couple of shirts, and a pair of shorts that I left for them on the back end of my trip, but they were thrilled all the same.

Even if it's just the smallest gesture, a word of kindness, or a drop of wisdom, give what you can and watch it blossom from there.

12

THE NEXT WIN

*There is no passion to be found playing
small—in settling for a life that is less than
the one you are capable of living.*

—NELSON MANDELA

In disgust, I threw my racket bag into the back of the hall closet and refused to even look at it for the next three months. I'd been working as a national coach for the USTA and I'd just come back from Albuquerque, New Mexico, where I'd been working with one of our young pros. While I was there, our boss put all coaches on a conference call with HR to inform us that we'd been fired as part of a reorganization.

"This is to inform you all that you will soon be terminated," said the impersonal voice on the other end of the line. This was our notice.

"What the . . . ?"

We could reapply for our positions later if we chose, without any guarantees, but that impersonal, obnoxious call would be our six-week final notice.

That was about all we could collectively grasp before the USTA bureaucrat hung up the phone. No room for further discussion or explanation. As much as I loved the young players I was coaching, and as natural as it felt to be in a hands-on teaching role, after this deflating experience I was done. We were spread out all over the world, away from home, our support systems, family, and friends. After all the hurt, confusion, and disappointment, we had to suck it up and see it through to the end, because we'd be doing a huge disservice to the athletes we were coaching if we distracted them with the details of what had just happened.

This was September 2002. From the age of six I had never been without a racket in my hand, but after twelve years of competing on the WTA Tour and going directly into this role for four and a half years of coaching, I needed a break. It was time to take stock, reflecting on all my experiences and ways I could channel them into a new career that could excite me as much as playing in a Grand Slam tournament.

Coaching was something I fell into naturally as I was phasing out of my professional career and dealing with an ankle injury. I could have kept going for a couple more years, but I knew I was no longer at my peak. By this time, I was ranked as high as 8th in the world in doubles, 67th in singles. Between 1988 and 1999, I'd reached the fourth round in singles at Wimbledon and taken twenty-one career Women's Tennis Association doubles titles playing alongside some of the greatest players in the world. Could I have achieved more? I certainly thought so. The one thing injuries do is help you play smarter to avoid reinjury and channel your energy more efficiently, although it's not necessarily worth the discomfort.

You must be honest with yourself and ask if you still feel as passionate about what you're doing as you did earlier in your career. Life on the tour was beginning to lose its luster. The harsh reality that professional athletes must face is that at some point, whether in our thirties or forties, the human body breaks down. We do all we can to stay in shape, train, and recover from strains, sprains, and broken bones. But at a certain point we must accept the fact that, while we can always play, we've reached the second set in our careers. Sooner in life than the average working adult, we have to start thinking about what comes next.

I'd already been sought after by the USTA to be a national coach. It was something that I had to think about deeply, because they wanted me to start in the middle of the Grand Slam tournament season in 1999. I didn't like the idea of stopping cold. I wanted to finish out the year. But I'd already accrued more than $1 million in winnings the minimum amount of money I'd promised myself I'd earn when I started out on the professional tour. It was short-sighted in retrospect, but at that time, it was a lot of money. I weighed my financial position with the opportunity I was being given and finally decided to get off the road, so I made a deal.

Although coaching was a ready-made career for me, I was losing the passion for competing. My main job was on the court with the kids and, as much as I loved them, the day-to-day physicality of the job was a grind. I spent my time off court honing my business skills, learning how to devise a budget, manage the travel and tournament schedules, balance the books, and keep up with files from Martin van Daalen, my immediate supervisor who was then based in Tampa. I also enlisted the help and guidance of some USTA Southern staff members in Atlanta, where I resided. I learned as I went along, then I decided I physically did not want to be on the court consistently like that. In fact, it felt like I was on the court more as a coach than I had been as a competitor.

With all these factors conspiring against my coaching career, I finally sat back and asked myself, "Okay, where do I want to go? What can I do?" I missed being excited to wake up every morning and tackle the next challenge with energy and purpose. I was still young, with a lot of working years ahead of me, and I needed to get that feeling back.

I always knew I wanted to commentate at some point, so I started putting myself in all the right places. I tried ESPN, but they were only interested in former Grand Slam champions. I always felt that by not using talent from different levels they missed out on the opportunity to diversify voices and perspectives. How lower-ranked players felt and operated was a story worth telling. Just because I wasn't a Grand Slam champion didn't mean that I didn't know the game and what was necessary to share with the viewers.

At an event in Los Angeles for the WTA Tour Finals in 2002 I started talking with people on the business and media side of the sport about what opportunities were out there and where I could fit in. The WTA didn't have as vast a staff as they have now. Being a tour supervisor would have been the only role available to me, but that wasn't what I wanted. Looking after the players and organizing the events and day-to-day activities that make tournaments run properly with the promoters was not something I was interested in doing. But having everyone there in one place was the perfect opportunity to network, so I sought out the advice of players who were already doing commentary, hoping that word would spread of my interest. Soon after, Pam Shriver called to let me know that the Tennis Channel was forming. Coincidentally, I knew the CEO at the time, and I made a call to arrange a meeting with the producer Larry Meyers and another executive.

For the first time since retiring from tennis I felt the thrill of anticipation. But now what? Although I'd studied communications at Northwestern University before leaving to go pro, my experience was limited. I had done some commentary work for Eurosport, a twenty-four-hour sports channel in Europe, and occasionally sat in as a guest analyst, although nothing extensive. But I had a plan of attack. This was in 2002, a time when Venus and Serena were No. 1 and No. 2 in the world, and we had mutual respect for each other. I researched the media landscape and asked myself, "What's missing? What's missing in the commentary world?" The answer to that question was clear: diversity.

"How can you not have a diverse analyst when your No. 1 and 2 players in the world are African American?" I challenged the people I'd been praying would hire me, hoping they wouldn't notice my hands were shaking. "I've played with and against them, I know them, I'm great friends with them, they're very difficult to get interviews with, and I know they're not going to turn me down if I ask them to do X, Y, and Z. Furthermore, I've had some skills in commentating—not a lot, but I'm a quick learner, and I'll do you proud. But you need someone who looks like me on your television screen when you come on air because that's really going to give you huge props right off the bat. In fact, you need me more than I need you!"

They bought it. Whew! And that's how my career off the court began.

We are not all destined to stay on one path. Many of us have two, three, or more careers in us. Although professional tennis had been my sole focus for most of my adult life, I did have other interests.

At Northwestern, I had focused on writing and public speaking, and enjoyed my coursework. It wasn't just some random major I'd picked out of a hat to allow me to engage all my energy on collegiate tennis. I was serious about the subject. I used to watch sports on television with my dad—everything from March Madness to *Monday Night Football* and, of course, the US Open and Wimbledon. Dad always made sure that I was watching sports on TV. He explained each sport and the differences in scoring and rules of each league, and I was fascinated. When I wasn't fantasizing about winning a Grand Slam, I daydreamed about what it might be like to interview athletes on network television or to sit behind that sports desk discussing that brilliant touchdown or bank shot.

As a player, I continued to be intrigued by what was happening on the other side of the camera. Whenever it was time to meet with the press after a match, I always made sure my hair, makeup, and outfits were on point. I wanted that extra polish to look good for the camera.

So commentating was always at the back of my mind. In fact, it had been one of my long-term goals from the time I was twelve.

A few days after my Tennis Channel interview, the network called and asked me for a sizzle reel. There were a few things on tape, but I had no idea where. So I asked my friend Sam Crenshaw, who was a sportscaster at the NBC affiliate in Atlanta, for advice. He invited me down to the station, instructing me to wear a nice suit. When I got there, he led me to the sports desk and started asking me questions about an Andre Agassi match during the Australian Open that he had playing on the video monitor. Confused, I played along. I figured he was just helping me get in some practice inside a real television studio. Before I knew what was happening, he said, "That's a wrap." Production had recorded our interaction and my commentary on what was on the monitor. I sent the tape to the Tennis Channel producers. Four months later, I got the call and was asked to be the color analyst for the upcoming Fed Cup tie in Lowell, Massachusetts, where the US team would host the Czech Republic. It would be my main role to assist the play-by-play commentator by filling in those moments with analysis when the play was not in progress. Tennis Channel was now live; mine was the first face anyone saw, alongside Barry Tompkins's.

I've seen how dramatically a match can go in a whole other direction just when you think you're in the final moments of a set, breathing new life into your playing. That's exactly what can happen in your career. Because most retire in their early to midthirties, tennis professionals are known for cultivating outside interests that could lead to something later in life. We have a lot of downtime between matches and tournaments to explore.

Often our outside pursuits are sports related, but not always. Earlier I mentioned how my successor Patrick Galbraith, chairman and president of the USTA, became a wealth manager after he retired from tennis. Then there is Helyn Edwards, my doubles and practice partner for adult tournaments when I was still a junior in Chicago. Her accomplishments are so varied and impressive, she's a modern-day Renaissance woman.

Helyn, who started playing tennis in 1959, remembers what it was like when she had to enter the courts of the country clubs through the kitchen. Not only was she a professional player and coach, teaching at the junior and collegiate level, but she also went on to obtain advanced degrees and certification in corrections psychology, which she teaches at the undergraduate level. She earned another degree in aviation and became a private pilot with single-engine, multi-engine, and instrument training. If that wasn't impressive enough, Helyn has also written multiple books and screenplays. She's also a Jim Crow–era and civil rights historian.

Helyn has been a member of my family for forty-four years. On winter weekends, when I was eight, she took me to the indoor courts at the Chicago Armory to practice, then stayed over at our house for dinner and a place to sleep because Mom didn't want her traveling far after dark. Helyn drove with us to tournaments all over the state. Being exposed to someone like Helyn throughout the course of my life and career was another reason why it never occurred to me to set career limits for myself. Fly planes, win matches, teach college courses, and write books? Why not?

Look around you. Seek out the people in your orbit who've managed to go big on their dreams despite their socioeconomic background. Observe those who have succeeded in more than one arena, whether that's a newly laid-off single mom making bow ties out of upcycled fabrics on YouTube or a corporate executive transitioning to a nonprofit because she wants to pursue something more meaningful than a six-figure paycheck. With rare exceptions, jobs for life are a thing of the past. We must embrace the changes and prepare to adapt, which often means having multiple careers during our working years.

One of my best friends growing up was a girl who lived down the block from us. Naomi Anderson (née Harvey) was raised by a single mom who spent her life working hard in factories so that her daughter wouldn't have to. Naomi and I practically lived in and out of each other's houses. While Mrs. Harvey was still at work, Mom would keep a watchful eye on us from the front room window as we ran around the neighborhood playing hopscotch or riding our bikes around the block.

Naomi had no interest in tennis, but she watched me closely

and paid attention to the ways my sport took me beyond the conventional path of college degree and safe day job. Naomi didn't exactly follow suit. She completed her degree at Loyola University and started working for a federal home loan bank while she was still in high school. But seeing me go for it inspired Naomi to push herself from one level to the next in her career in back-office finance, whether that was at a prepaid card company or an auto loan processing center where she worked as vice president.

"Seeing you from across the fence made me say to myself, 'Yes, I can,'" Naomi told me recently. "With each progression of my career, I can see how I'm going to succeed. As two little kids from Chicago's West Side, we've done okay!"

I was enjoying my new career in sports commentating, which had advanced to include regular spots on the first all-female sports roundtable show, CBS Sports Network's *We Need to Talk*. But something was still missing. That need to give back not just to my sport but to my community was driving me. That feeling that there was more for me to accomplish, combined with my self-belief, was what led me to my leadership role with the HJTEP and, ultimately, the USTA. All these experiences brought me to where I am now—a place of influence in all the major arenas of my sport. Today I am uniquely positioned to make an impact on all the issues I care about.

You never know how long the set you're playing is going to last. Knowing that I ultimately wanted to broaden my experience beyond the tennis world, my two-year term extension with the USTA allowed me to continue to solidify relationships with corporate sponsors, giving me more exposure and keeping me in the same orbit as some of the best minds in corporate and nonprofit leadership.

Through playing tennis and operating at all levels of the tennis world, I developed traits that would translate well to business: perseverance, discipline, and a competitive spirit. A sport that, more than most, requires the honing of your mental game, tennis also gave me self-accountability and an ability to strategize. All the finer points of management, whether of myself as I was navigating life or of my team within an organization, came into play.

There are so many parallels between the sport and an individual's path to success, from the obvious work ethic it takes to practice, persist, and excel to the sportsmanship that elevates tennis, and everyone involved, including the millions who watch it. As a player I could see the whole court and work within its boundaries; correspondingly, as a leader I understood my industry from all sides of the prism and could operate effectively within its system. Another two years at the helm of the USTA would help me to hone and polish my executive skills as I prepared to step into an entirely new arena. Hitting a ball on those cement courts in Garfield Park all those years ago, who knew it would come to this?

Later in life I also developed a passion for golf. The sport has been key to my building an extended network of relationships. And talk about being the only one! A black woman on a golf course was and still is a rare sight indeed. But that was my advantage. People remembered me and enjoyed the fact that we share a love for the game despite our outward differences. I've had so many illuminating conversations with the—I'll just say it—mostly old white men I've met on the links. Many were captains of industry who were more than happy to talk about their businesses and share their insights. These encounters have expanded my world in no small measure.

Never limit your own possibilities. When your parents, teachers, or coaches said, "You can do anything," it wasn't just something to say. It was the truth. When you put the work in while staying open to the opportunities that come your way, you never know what lies ahead. Sometimes it's bigger than anything you can dream up.

You might have spent years in school getting a professional degree, working internships, specializing in certain areas of your industry, and following a prescribed path for someone with your academic and work credentials. If that's what fulfills you, fantastic! Keep going! But if you're curious about what else may be out there, give yourself permission to explore. Try out internships in other areas. The world is wide open to you, and your journey through it can be more satisfying when you take an alternate route.

I was reminded that our potential is limitless early in 2019, soon after my tenure ended from leadership of the USTA—another huge career transition. I was invited to the Red Bull Air Force Fly Girls Summit in Orlando to sit on a panel and discuss the role of women in sports. I immediately called Roberta and asked her if she wanted to go because skydiving was on her bucket list. One of the fascinating things I learned was that skydiving is one of the most male-dominated sports around: eighty-six female tandem instructors compared with almost two thousand male instructors.

During the two-day event, which was hosted by Amy Chmelecki, the skydiving champion of 2018, and included guests like Melissa Arnot Reid, the first woman to successfully summit Mount Everest, and Grete Eliassen, six-time Winter X Games gold medalist in freeskiing and Women's Sports Foundation past president, we discussed issues like increased participation, equal pay and recognition, and the importance of assuming leadership roles in their respective sports. "When you want to do your first skydive in the United States you literally have to strap yourself to a man to enter the sport, so we are here to change that experience," Amy, who used to spend all her money from bartending tips on jumps, told us.

After the panel, I was invited to jump out of a plane. "Why not?" I figured. What better way to enter the fourth set of my career than through a skydive with some badass women?

Sliding out of that plane was at once terrifying and exhilarating. As we flew 13,000 feet above the orange groves, strip malls, and amusement parks, I felt like Superwoman.

It was all I needed to make the next leap: faith, a parachute, and a world-class female tandem instructor.

MY WORDS WERE MISCONSTRUED

People can misunderstand anger for strength because they can't differentiate between the two. No one has stood up for themselves the way you have, and you need to continue trailblazing.

—NAOMI OSAKA IN A TEXT MESSAGE
TO SERENA WILLIAMS

There was plenty to savor during the Sunday of the 2018 men's final, the last time that I would preside over a Grand Slam tournament. Earlier that afternoon, international delegates from all over the world greeted me in the President's Suite Dining Room. The Grand Slam tournament chairmen each presented me with a moving toast and tribute. My colleagues in the world of tennis made heartfelt speeches about what I'd meant to them and the organization. This was done in front of my family members and dearest friends. It was a gift, one of those rare opportunities to

be with people you love and respect most in the world and hear all
that you've tried to achieve played back to you in their own kind
words. I should have soaked it all up, but the day after the drama
of the women's final I was nowhere near as present to the moment
as I would like to have been.

It didn't help that tennis bad boy John McEnroe was going off
script. My team at the USTA had gone back and forth with John's
handlers several times about what was supposed to happen on the
podium at the 2018 US Open Men's Final. It should have been
straightforward. He was to hand the trophy to me; I would then
present it to the winner, Novak Djokovic. Yet in the days preced-
ing, there'd been a flurry of requests for "clarification." Apparently,
it wasn't clear enough, because when the moment came, John took
the trophy from the stand and proceeded to walk right past me to
present it to the champion himself.

Fortunately, Novak, a fifteen-time Grand Slam tournament
singles champion for whom this had by now become a familiar
drill, also noticed the presentation going sideways. On some level
the mishap was understandable, as this was the first year that we
had former champions on the court during the trophy ceremony.
Like the true pro he is, Novak graciously accepted the trophy from
John, then, knowing full well how the scene was supposed to play
out, slid over to where I was standing to kiss me on both cheeks
and thank me, with an amused twinkle in his eye. He'd made it
seem to the millions of viewers and spectators in the stands like
it was all part of the plan. Crisis averted. This was my last official
duty with the world watching.

Count on John to do it his way. How could we have possibly expected anything else? Put a shiny silver trophy in front of
the most pathologically competitive player on the planet, and of
course he's going to find it hard to let go! A few years earlier, he

played a marquee match in Madison Square Garden with my tennis dad, David Dinkins, for a fundraiser. John was supposed to let him win, but he couldn't bring himself to throw a game, so he kept firing ace shots at His Honor, who was pushing eighty at the time. We can laugh about it now, but poor David looked like he was out there dodging bullets.

As was I. In the aftermath of the Naomi and Serena match, everyone from respected pundits to social media trolls had gotten hold of this story, tearing it to pieces like a rabid dog. The world's press continued to hound me.

But the fuss eventually died down, moving on to the next news cycle. Like Serena and Naomi, I found peace after the women's finals maelstrom. I had done what mattered most by addressing the situation with those who were most affected. Now I was ready to close the book on all the drama.

In a near perfect circle moment, Naomi Osaka demonstrated the same compassion and sportsmanship toward her opponent at the US Open in 2019 as Serena had served up following their infamous women's finals match in 2018.

Naomi was playing against fifteen-year-old Coco Gauff, a native of Atlanta, Georgia, who had become a household name after defeating her idol, Venus Williams, in the opening round of Wimbledon, then shocked the world by reaching the fourth round. She followed that performance by winning the doubles title with fellow American Caty McNally at the Citi Open in Washington, DC, before playing in the US Open. These performances gave her the confidence that she took into the fall season, winning her first WTA singles title at the Linz Open, making her the youngest singles title holder on the WTA Tour since Maria Sharapova won Wimbledon in 2004 at seventeen. Coco followed that up by winning her second doubles title with McNally.

When she finally lost all that momentum from earlier in the Grand Slam season, losing to Naomi 6-3, 6-0 in the third round, Coco was devastated and could not contain her tears. Seeing this, twenty-one-year-old Naomi, now savvy and seasoned about these situations, walked over to console the young phenom, telling her it was okay to cry on the court, an experience she knew all too well. Then she gently insisted that Coco join her in the post-match interview with ESPN's Mary Joe Fernández, a spot usually only reserved for the winner:

> "She did amazing, and I'm going to learn a lot from this match," a choked-up Coco told the crowd. Then she turned to Naomi and said, "She's been so sweet to me, so thank you for this. Thank you."

Addressing Coco's mom and dad in the stands, Naomi said, "I remember I used to see you guys training in the same place as us, and, for me, the fact that both of us made it, and we're both still working as hard as we can, it's incredible. I think you guys are amazing, and Coco, I think *you're* amazing."

There wasn't a dry eye in the stadium.

That moment put it all into perspective. My biggest hope is that people will look back on my career and say, "She embraced all." I think I've been a voice for everyone, especially women. It never gets old when someone stops me on the street—minorities, women, and others—just to say, "Thank you for making me feel like there's a place for me in this sport; thank you for making me feel welcome."

"No," I say. "Thank *you* for embracing me and allowing me to own my arena, wherever I go and wherever I speak."

And thank you for the opportunity to inspire you to own *your* arena.

ACKNOWLEDGMENTS

The signature quote at the end of my emails reads: "Embrace the path that you lead and enjoy the battle." There have been many who have joined me on this path, without whose support I would not be where I am today. So many, in fact, that my publisher may run out of ink, but here it goes . . .

At the top of this list is my family: my beloved parents James and Yvonne Adams. Throughout all my years of tennis, I never came across a stronger doubles team. To my mother and father, whose loving guidance and impact allowed me to "own the arena." You sacrificed to support your belief in me, and you encouraged me to walk confidently down an uncharted path, giving me the confidence to unabashedly enter any space. You enforced and emphasized the importance of education and being a leader. You taught me the value of hard work and perseverance in order to rise to the top. And to my brothers Myron, Maurice Adams, and Victor Bruce, and my nieces Andrea Guidry and Azure Bruce.

And to Roberta Graves. I would not have excelled or risen to this level in my executive career without her as the constant buzzing bee in my ear, letting me know I could always be more and always encouraging me to own the arena.

Next comes my extended tennis family, the folks who were integral to my development as a player: Joe Breedlove, Sandy Stap Clifton, Charlton Eagle, Helyn Edwards, Tony Fox, Patricia Freebody, John Lucas Jr., Dave Muir, Tex Richardson, Jeff

Rothstein, Christopher Scott, Rod Schroeder, Charles Searles, Jane Sieffert, Bill Simms, Willis Thomas, John Wilkerson, Kim Williams, Moses Vincent, Joe White, and Donna Yuritic.

My doubles partners: Manon Bollegraf, Mariaan de Swardt, Zina Garrison, Gail Gibson, Penny Mager, Lori McNeil, Lisa Pamintuan, Chanda Rubin, Pam Shriver, Debbie Graham Shaffer, Diane Donnelly Stone, the Northwestern University tennis team (1985–1987), and the Whitney Young High School tennis team (1981–1984).

I've also been blessed with an outstanding team of mentors: David Grain, Dr. Lisa Grain, Earl "Butch" Graves Jr., Jane Brown Grimes, David Haggerty, Jay Hans, Peachy Kellmeyer, James R. Kelly III, Billie Jean King, Alan Schwartz, Judy Smith, Kathleen Stroia, Don Tisdel, and Jon Vegosen.

And a long list of friends and supporters: Jeff Appel, Stacey Allaster, David and Sandra Bishop, Everick and Lisa Brown, Christopher Clouse, Kathy Francis, Andrea Hirsch, Ilana Kloss, Charlotte Kuey, Judy Levering, Lauren Lowery, Ndidi Massay, Gary Mouton, Mark Preston, Lily Kelly Radford, Lori Shaw, Gordon Smith, Dr. Janet Taylor, Vanessa L. Williams, and Wendy Willis.

I'd also like to thank my supporters at the Chicago Prairie Tennis Club: Hazel Bond, Melba and Richard Bradley, Joyce Clark, Beverly Coleman, Angie Grant, Ruth Leake, Nancy and Ron Mitchell, Julia Steele, and Helen Watson.

Thanks to my tennis big sisters: Pam Black, Liza Cruzat, Kendall Dogan, Donna Douglas, Ruth Hadnot, Julie Jones, Carol Kay, Stacey Knowles, Leslie Pilot, Lisa Thomas, and Pam Triche.

And to my tennis big brothers: Edward Cruzat, Darryl Dogan, Sylvester Dorsey, Ivan Dixon, Rev. Marty Goole, Jeff Gordon,

Tyrone Mason, Mel Phillips, Darryl Pope, Eric Smith, Tim Tyler, and Donald Young Sr.

I am grateful to the teachers who have shaped me: Joyce Baker, Joyce Myers, Emma Teagues, and professor Robert Gundlach.

I am grateful to the individuals who worked alongside me and pushed me to be my best: Ricardo Acuña, Andy Andrews, Patrick Galbraith, Rodney Harmon, Mikella Matthias, Nellie Nevarez, Martin van Daalen, and Carol Watson.

And my extended family who supported me: Bernard Clay, the Adams family, the Aitch family, the Barry family, the Garrison family, the McJunkins family, the Rogers family, the Thomas family, and the Smith family.

Too numerous to mention are: the Harlem Junior Tennis and Education Program board members, HJTEP families, HJTEP staff, and HJTEP supporters.

As well as the teams at: CBS Sports Network, International Tennis Federation, International Tennis Hall of Fame, Tennis Channel, United States Tennis Association, USTA Foundation, *We Need to Talk* team, and World Team Tennis.

Last but by no means least, I'd like to thank my editorial team for their invaluable contributions to the making of this book, including my publisher, Judith Curr, director of publicity Paul Olsewski, and the entire HarperOne group. In particular I would like to thank editor Tracy Sherrod, who gently pushed me to dig deeper; my literary agent, Carol Mann, who knew there was a book in me long before I did; and my tireless collaborator, Samantha Marshall, who understood the story I wanted to tell and worked with me through many long days to help me produce something that I am proud to share with the world.

ABOUT THE AUTHOR

After serving an unprecedented two consecutive terms as USTA chairman and president, Katrina M. Adams began serving a two-year term as immediate past president in January 2019.

Adams was the first African American, first former professional tennis player, and youngest person ever to serve as USTA president. During her administration, the USTA achieved a number of major milestones, including the opening of the USTA National Campus in Orlando, Florida, and the completion of the strategic transformation of the USTA Billie Jean King National Tennis Center in Flushing, New York. She also spearheaded an unprecedented outreach effort into underserved communities—with an emphasis on Latino communities—in an effort to share the sport of tennis with more people.

An accomplished professional player, Adams competed for twelve years on the WTA Tour. She ranked as high as No. 67 in the world in singles and No. 8 in doubles, winning twenty doubles titles and reaching the quarterfinals or better in doubles at all four Grand Slam tournaments. Her best Grand Slam singles result was reaching the fourth round at Wimbledon in 1988. She attended and played collegiate tennis at Northwestern University, majoring in communications and helping the Wildcats to a Big Ten Conference championship in 1986. That year, she was the Intercollegiate Tennis Association (ITA) Rookie of the Year. In 1986 and 1987, Adams was named NCAA All-American. Also in 1987, she became the first African American to win the NCAA doubles title. In 1997, she was inducted into the Northwestern University Athletic Hall of Fame.

While an active pro, Adams served on the board of directors of the WTA as a player representative for four one-year terms and on the WTA's Player Association for five two-year terms. After leaving the tour, Adams was a USTA National Tennis Coach from 1999 to 2002. She joined the USTA Board of Directors in 2005, serving as a director at large and as the association's vice president and first vice president before assuming the presidency.

Adams was elected the vice president of the International Tennis Federation in 2015 and reelected in 2019. She has served as chairman of the Fed Cup Committee since 2016. She was named chairman of the Gender Equality in Tennis Committee in 2018 and in 2020 was appointed to the Finance and Audit Committee. She was also named to *Adweek* magazine's "Most Powerful Women in Sports" list in 2016 and 2017, and *Forbes* magazine's "Most Powerful Women in Sports" list in 2017. Also in 2017, Adams was named to *Ebony* magazine's "Power 100" list. In 2018, Adams became a member of the board of directors of Pivotal Acquisition Corp., in 2019 joined Pivotal Acquisition Corp. II, an emerging growth company, and in 2020 joined the Athletes Unlimited Advisory Board.

Since 2005, Adams has served as the executive director of the Harlem Junior Tennis and Education Program, based in New York City. A Chicago native, Adams currently lives in Yonkers, New York.